WASHINGTON SUMMIT PUBLISHERS
2020

Also by Edward Dutton

Meeting Jesus at University: Rites of Passage and Student Evangelicals (2008)

The Finnuit: Finnish Culture and the Religion of Uniqueness (2009)

Culture Shock and Multiculturalism (2012)

Religion and Intelligence: An Evolutionary Analysis (2014)

The Ruler of Cheshire: Sir Piers Dutton, Tudor Gangland and the Violent Politics of the Palatine (2015)

How to Judge People By What They Look Like (2018)

J. Philippe Rushton: A Life History Perspective (2018)

The Silent Rape Epidemic: How the Finns Were Groomed to Love their Abusers (2019)

Race Differences in Ethnocentrism (2019)

Churchill's Headmaster: The "Sadist" Who Nearly Saved the British Empire (2019)

With Bruce Charlton

The Genius Famine: Why We Need Geniuses, Why They're Dying Out and Why We Must Rescue Them (2016)

With Richard Lynn

Race and Sport: Evolution and Racial Differences in Sporting Ability (2015)

With Michael Woodley of Menie

At Our Wits' End: Why We're Becoming Less Intelligent and What it Means for the Future (2018)

Islam
An Evolutionary Approach

Edward Dutton

© 2020 by Edward Dutton
All rights reserved.

No part of this publication may be reproduced, distributed, or transmitted in any form or by any means, including photocopying, recording, or other electronic or mechanical methods, or by any information storage and retrieval system, without prior written permission from the publisher, except for brief quotations embodied in critical reviews and certain other non-commercial uses permitted by copyright law. For permission requests, contact the publisher.

Washington Summit Publishers
P.O. Box 1676
Whitefish, MT 59937

email : hello@WashSummit.com
web: www.WashSummit.com

Cataloging-in-Publication Data is on file with the Library of Congress

ISBN: 978-1-59368-063-3
eISBN: 978-1-59368-064-0

Printed in the United States of America
10 9 8 7 6 5 4 3 2 1

CONTENTS

Acknowledgments

Parts of this book, in slightly different forms, have previously appeared in *At Our Wits' End: Why We're Becoming Less Intelligent and What It Means for the Future* (2018), which I co-authored with Michael Woodley of Menie, and in "The Mutant Says in His Heart, 'There Is No God': The Rejection of Collective Religiosity Centred Around the Worship of Moral Gods is Associated with High Mutational Load" (2018), which appeared in *Evolutionary Psychological Science* and which I co-wrote with Guy Madison, and Curtis Dunkel. I acknowledge my intellectual debt to these colleagues. I would also like to thank Herr Emil Kirkegaard for drawing my attention to some highly relevant literature, as well as Richard Spencer and Nils Wegner for editing and developing this book.

It is appreciated that this is a rather "controversial" topic and it cannot be emphasized enough that there are many positive dimensions to Islam; indeed, it is how these positive dimensions manifest themselves that has fascinated me for so long. But at the same time, there is no reason why we shouldn't be clear what we mean. I first became interested in Islam in about 1987, because my best friends at Infants' School were Muslim. Accordingly, I would like to dedicate this book to Yusuf and Sa'ad, as well as to their respective parents and siblings.

Edward Dutton
December 2019
Oulu, Finland

Preface

John Derbyshire

As we come into the third decade of the 21st century, the human sciences are in a peculiar state.

Good quantitative knowledge about human nature—good enough for the kind of robust speculation that energizes scientific discovery, sometimes good enough to have immediate predictive value—is accumulating very fast. Yet even as our understanding advances, public discussion of topics in the human sciences is fiercely constrained by obscurantist political taboos.

Indeed, as knowledge advances, the constraints seem to tighten, the taboos to multiply. When, at the turn of the century, earnest editorialists in respectable broadsheet newspapers were assuring us that "There Is No Such Thing as Race," I thought we had reached peak obscurantism, and that the taboos could surely not stand much longer on such patently false foundations. Now, 20 years on, those editorialists are telling us that "There Is No Such Thing as Sex." What next? There Is No Such Thing As Age? No Such Things As Day and Night?

Fortunately, the armies of obscurantist orthodoxy don't have things all their own way. A small defiant cadre of human-science researchers have been laboring to bring our new understandings to general attention. Indeed, we now have a new generation of these dissidents. David Becker, Davide Piffer, and Michael Woodley of Menie are following the trails blazed by Hans Eysenck, Arthur Jensen, Richard Lynn, Helmuth Nyborg, Bruce Charlton, J. Philippe Rushton, and Tatu Vanhanen.

Edward Dutton is one of the liveliest and most engaging of this new generation of academic dissidents. His YouTube channel, *The Jolly Heretic*, presents entertaining commentary informed by wide and deep reading in evolutionary psychology and anthropology. Edward Dutton is what Bill Nye the Science Guy would be, if that gentleman dared to present the human sciences with uninhibited objectivity. Dutton is particularly good on the subject of religion, the topic of his Ph.D. thesis. It was inevitable that he would sooner or later give us a book on Islam, and here it is.

Why do Muslim nations have lower mean IQ than non-Muslim nations they are genetically close to? How does religiosity relate to ethnocentrism? Does the custom of purdah increase paternal investment? ("Dads not cads"). Does it cause vitamin D deficiency in women? Does a month of compulsory fasting affect pregnancy? Why did Muslim creativity decline after A.D. 1000? How does slavery affect the gene pool? Our author offers reasoned, well-informed arguments on these and a host of other questions we have all wondered about at some time or other.

And what are the prospects for Western societies being transformed by mass Muslim immigration? Shall our flagging civilizational energies be reinvigorated by a new infusion of religiosity? Or are we doomed to inherit those "improvident habits . . . and insecurity of property" that Winston Churchill says "exist wherever the followers of the Prophet rule or live"? We shall find out. Until we do, here is Edward Dutton to weight the probabilities for us from a deep, solid base of learning and insight.

Huntington, New York
December 10, 2019

Islam

An Evolutionary Approach

Chapter 1

Improvident and Sluggish

How dreadful are the curses which Mohammedanism lays on its votaries!
—Winston Churchill

Muslims and IQ

Winston Churchill summed up Islam in 1898 in his book *The Story of the Malakand Field Force.* He described the British army's campaign, of which he was part, in what is now the border area between Afghanistan and Pakistan, passing comment on the culture of the natives as he did so:

> *How dreadful are the curses which Mohammedanism lays on its votaries! Besides the fanatical frenzy, which is as dangerous in a man as hydrophobia in a dog, there is this fearful fatalistic apathy. The effects are apparent in many countries. Improvident habits, slovenly systems of agriculture, sluggish methods of commerce, and*

> *insecurity of property exist wherever the followers of the Prophet rule or live.*[1]

Cutting through the flowery Victorian language, what Churchill meant was that, when it comes to Muslim peoples, there is, on average, not that much going on upstairs: They aren't very intelligent, and certainly they are nowhere near as intelligent as Europeans. This is simply an empirical fact.

In his book *Race Differences in Intelligence*, British psychologist Richard Lynn brings together a large number of samples to estimate the average IQ of the North African and Middle Eastern countries; these being overwhelmingly Muslim.[2] It is 84, compared to a Western European IQ of 100. In other words, the average IQ in Muslim countries is the same as that of the average IQ of a low-level security guard or casual laborer in a Western country.[3] As we will see shortly in more detail, these 16 points make a huge impact. They are the difference between the average secondary or high school science teacher and the average science professor at a university; or between the average Western person—who sits behind a computer all day, much like in the TV sitcom *The Office*—and the average high school science teacher.

1—Winston Leonard Spencer-Churchill, *The Story of the Malakand Field Force: An Episode of Frontier War* (London: Dover Publications, 1898), Chapter X.

2—Richard Lynn, *Race Differences in Intelligence: An Evolutionary Analysis* (Whitefish, Montana: Washington Summit Publishers, 2015 [2005].

3—See also Richard Herrnstein and Charles Murray, *The Bell Curve: Intelligence and Class Structure in American Life* (New York: Free Press, 1994).

This low average IQ can also be found among the increasing number of Muslim immigrants, and their descendants, living in Western countries. Danish males are subject to a system of military draft. Danish researcher Emil Kirkegaard has drawn upon published Danish army conscript data to show that if we set the average IQ of Danes at 100 then the average IQ of non-Western immigrants in Denmark, most of whom are Muslim, is roughly 86.[4] He cites studies showing that in many different Western countries, a comparable difference exists between the native population and immigrants and their offspring, at least when putting aside immigrants from Northeast Asia: the average IQ of the Chinese is 104.[5] Kirkegaard demonstrates that this difference is substantially genetic in origin, because it is on the parts of the IQ test, the results of which are the most strongly genetic (see "What is Intelligence?" below). This is another empirical fact. The earlier a trait is noticeable, the more likely it is to be mainly genetic. In England, for example, race differences in IQ have been shown to develop by a very young age among second-generation non-Western immigrants, the difference fits with studies that show average differences in IQ between countries, and it is further consistent with evidence that the frequency of gene-forms associated with very high intelligence is lower in non-Western than

4—Emil Kirkegaard, "Predicting Immigrant IQ from their Countries of Origin, and Lynn's National IQs: A Case Study from Denmark," *Mankind Quarterly* 54 (2013): 151-167.
5—Richard Lynn and David Becker, *The Intelligence of Nations* (London: Ulster Institute for Social Research), 67

in Western countries.[6] The key point is that immigrants and their descendants in Denmark are overwhelmingly Muslim, so what Kirkegaard is effectively finding is that Muslims in Western countries have IQs which are almost 15 points lower than the native population's IQ.

Fascinatingly, low IQ immigration from the Indian Sub-Continent is very much a Muslim issue, rather than, as some might believe, an issue solely of "race" and being genetically "South Asian." According to UK data from 2006, when we set the average IQ of native British people at 100, then the average IQ of people of Chinese ancestry is 103 and the average IQ of those of Indian ancestry—people who are overwhelmingly Hindu—is 95. However, among those whose ethnic origins are in Pakistan or Bangladesh— where people are overwhelmingly Muslim—the average IQ is 92. Worryingly, in the UK, the White British have an average of 1.7 children; the Indians, an average of 2.2, and the Muslims, 5.[7]

This should be deeply concerning when you consider that a meta-analysis of twin studies—which allow you to test for how genetic a trait is—has shown that IQ is about 0.8 genetic and thus overwhelmingly inherited from one's parents.[8] The key environmental component to intelligence, apart from nutrition at a young age, appears to be an intellectually stimulating

6—Davide Piffer, "Correlation Between PGS and Environmental Variables," *RPubs*, https://rpubs.com/Daxide/377423 (accessed August 15, 2019).
7—Richard Lynn, *Dysgenics: Genetic Deterioration in Modern Populations, Second* Edition (London: Ulster Institute for Social Research), 269.
8—*Ibid.*, 101.

environment.[9] This is especially important during key growth phases, but it is also true throughout your life. The brain is a muscle and it can gain mass, up to a phenotypic limit, but it can also lose it. If you spend your time with extremely intelligent people, generally stimulating your mind, then your intelligence will be pushed to its maximum phenotypic limit. If you have an intellectually stimulating job—as a research scientist or a lawyer or a doctor, for example—but you then retire and abandon these areas completely, dedicating yourself to golf and your grandchildren, then your intelligence will decline.[10]

Accordingly, there really does appear to be very little question about it. Muslims are, when compared to those who are from countries that are genetically relatively similar but dominated by other religions, not very bright; something that has been shown to lead, as Churchill observed, to "apathy" and being "slovenly," "improvident," "sluggish." But Churchill doesn't really answer the question "Why?" Why are Muslims so deficient in intelligence? Is it that people of low intelligence tend to be less likely to reject Islam, if that is the religion in which they have been raised? Or is it that Islam actually *makes you stupid*? Are there things about the nature of Islam that somehow create people who just aren't very clever? And if that is so,

9—Anett Nyaradi, Jianghong Li, Siobhan Hickling, Jonathan Foster, and Wendy H. Oddy, "The Role of Nutrition in Children's Neurocognitive Development, From Pregnancy Through Childhood," *Frontiers in Human Neuroscience* 7 (2013): 97, doi:10.3389/fnhum.2013.00097.

10—See James R. Flynn, *Does Your Family Make You Smarter? Nature, Nurture and Human Autonomy* (Cambridge: Cambridge University Press, 2016).

what are those things? And how can Muslim countries have even survived in battle against other—presumably more militarily strategic—peoples, if they are, apparently, so intellectually backward?

In this book, I will explore the ways in which Islam does indeed make you stupid, but I will further show how, by doing this, it, ironically, allows Muslims to gradually dominate and displace far more intelligent peoples. And this leads us to another cuttingly accurate remark that Churchill made about Islam, again in *The Story of the Malakand Field Force*:

> *But the Mahommedan religion increases, instead of lessening, the fury of intolerance. It was originally propagated by the sword, and ever since, its votaries have been subject, above the people of all other creeds, to this form of madness. In a moment the fruits of patient toil, the prospects of material prosperity, the fear of death itself, are flung aside.*[11]

In other words, Muslims are prepared to make extraordinary sacrifices for their faith, fanatically confident in the words of the Prophet Muhammad (c.570-632) and the power of Allah. Islam reduces your intelligence, but, by doing so, we will see that it increases the extent to which you are high in positive and negative ethnocentrism—the extent to which you are prepared to make sacrifices for your group and obliterate those who challenge it.

In this book, we will see that almost all of the Islamic doctrines that tend to reduce your intelligence also

11—Churchill, *The Story of the Malakand Field Force*, *op. cit.*, Chapter III.

increase the extent of your ethnocentrism and thus the ethnocentrism of your group. And computer models, as we will observe, are quite clear: the more ethnocentric group—the one whose members are cooperative with and make sacrifices for each other but who shun or actively attack outsiders—will always come to dominate and wipe out groups who are altruistic to everyone.

In real life, however, this "ethnocentric strategy" can only go so far. Islam will sweep away all before it until it meets a group whose intelligence is so much higher than its own that this high IQ more than compensates for what the group lacks in ethnocentrism. The group's relative lack of ethnocentrism will mean that it will be better able to trade, gain new knowledge from other groups, expand and so increase its gene pool. It will make it open to potentially useful foreign ideas and curious about new things, as is specifically the case when we compare Europeans to other races, such as the Northeast Asians.[12] Such a high intelligence group will have produced outliers, blessed, just by genetic chance, with fantastically high intelligence combined with the personality traits of moderately low Agreeableness (altruism) and moderately low Conscientiousness (rule-following and impulse control).

This unlikely psychological profile—rendered more likely by this relatively open group's larger gene pool—allows those who have it, genius, to come up with highly original and useful ideas because it means that

12—Kenya Kura, Jan te Nijenhuis, Edward Dutton, "Why Do Northeast Asians Win So Few Nobel Prizes?" *Comprehensive Psychology* 4 (2015): https://journals.sagepub.com/doi/10.2466/04.17.CP.4.15 (accessed August 15, 2019).

they can 'think outside the box' and they don't really care about offending people, with original ideas almost always offending vested interests.[13]

These "geniuses" will have originated brilliant weapons, or they will be superb tactical generals.[14] And this "genius strategy" will often be enough to overpower even the most ethnocentric of opponents, as long as the genius group maintains an optimum relatively low level of ethnocentrism.[15] As we will see, this is what Islam was confronted with in the form of the European imperial powers in the eighteenth and nineteenth centuries. But the ethnocentrism levels of those peoples—the Western Europeans—have now dropped so low, for reasons we will explore, that the Islamic strategy is now working even against *them*. This reversal might have started as far back as 1885, when Sudan's infamous *Mahdi,* Muhammed Ahmad (1844-1885), managed to take over Sudan and slaughter British-led forces. His fanatical army killed Sudan's governor, General Gordon of Khartoum (1833-1885), paraded his head around Khartoum, and established a Caliphate which was not overthrown until 1899.[16]

13—Michael A. Woodley of Menie and Aurelio José Figueredo, *Historical Variability in Heritable General Intelligence: Its Evolutionary Origins and Socio-Cultural Consequences* (Buckingham, UK: University of Buckingham Press, 2013).

14—Edward Dutton, Bruce G. Charlton, *The Genius Famine: Why We Need Geniuses, Why They're Dying Out and Why We Must Rescue Them* (Buckingham: University of Buckingham Press, 2015).

15—See Edward Dutton, *Race Differences in Ethnocentrism* (London: Arktos, 2019).

16—Fergus Nicholl, *The Sword of the Prophet: The Mahdi of Sudan and the Death of General Gordon* (Stroud, Glos: The History Press, 2004).

Islam is conquering Western Europe. The evidence for this has been set out in considerable detail, with the focus on France, in the book *The Colonisation of Europe* by French philosopher Guillaume Faye (1949-2019). Muslim countries, that were once colonized by France's military, are now quite deliberately themselves colonizing France with the wombs of their women, expanding because Western colonization of their countries has permitted them to expand. As Guillaume Faye puts it:

> *Colonialism has turned on us like a boomerang . . . By offering our medical techniques we have lowered their mortality rate and caused an explosion in their demographics at our own expense.*[17]

Muslim-world politicians, such as in Algeria, have openly advocated this process. It is a "settlement from below" where the Islamic world exploits Western European charitable impulses to colonize Western European countries.[18] Due to the high fertility of Muslims in France, they are gradually bringing more and more areas under *de facto* Islamic rule and wielding their influence nationwide. They appear to be relatively integrated immigrants, argues Faye, until a tipping point is reached, after which they begin to assert themselves. And Western Europe's ethnocentrism has dropped so low, in part because of the cozy environment its intelligence has created, that it has simply become too nice, unable to deal with the optimum combination of stupidity and

17—Guillaume Faye, *The Colonisation of Europe*, trans. Roger Adwan (Budapest: Arktos, 2016), 13.
18—*Ibid.*, 12, 54.

ethnocentrism which Islam represents. However, before we begin there are two key concepts that we have to get our heads around: Intelligence and Group Selection.

What is Intelligence?

"Intelligence" is defined as "the ability to reason, plan, solve problems, think abstractly, comprehend complex ideas, learn quickly, and learn from experience."[19] In other words, it is the ability to solve complex problems and solve them quickly. The quicker you can solve the problem, the more intelligent you are, and the harder the problem has to be before it is simply beyond you then the more intelligent you are. Intelligence is quantified by the intelligence quotient (IQ), a score that can be derived from a number of standardized tests. These tests measure reasoning ability across a wide range of domains, for example verbal, mathematical, and spatial. Ability in each of these subsections positively correlates with ability in the other subsections, allowing us to posit that they together measure an underlying factor called "general intelligence" or *g*. This is the essence of intelligence, and it is generally what we are talking about when we say that one person is more "intelligent" than another. A person's IQ score, therefore, is not the same as their "general intelligence." The IQ test is an imperfect measure of *g*. It does measure this but also measures so-called "specialized abilities" which correlate with *g*. Different parts of the IQ test are

19—Linda Gottfredson, "Editorial: Mainstream Science on Intelligence," *Intelligence* 24 (1997): 13.

more, or less, "*g*-loaded" than others. For example, the linguistic component will tend to be less *g*-loaded than the mathematical component. A correlation, it should be noted, refers to the relationship between two variables or the extent to which one predicts the other. This can be either a positive or negative relationship. A correlation of 1 means that one variable always predicts the other. "Statistical significance" is how scientists test, using calculations based on the strength of the correlation and the size of the sample, whether or not the correlation is merely a fluke. It is accepted, based on this, that if we can be at least 95 percent certain it is not a fluke then the relationship is statistically significant ($p \leq 0.05$) and thus *real*.

It cannot possibly be argued that intelligence is somehow "a very Western concept," something which a female Finnish cultural anthropologist once told me, with great profundity, at a Halloween Party in 2014. Proxies for intelligence include general knowledge, something which is valued in all cultures.[20] Intelligence is negatively associated with criminality, which is disliked in all cultures. It also robustly positively correlates with education, income, and health. As such, intelligence cannot be dismissed as only relevant in the West, nor can it be dismissed as unimportant. Intelligence also cannot be argued to be "too complex to understand," because, as science is in a constant state of progress towards greater understanding, this could be argued to be true of anything at all. A full list of the traits associated with intelligence is laid out in Table 1.

20—David M. Buss, *The Evolution of Desire: Strategies of Human Mating* (New York: Basic Books, 1994).

Table 1. Behaviors and Preferences Associated with Intelligence[21]

POSITIVE CORRELATION	NEGATIVE CORRELATION
Achievement motivation	Accident proneness
Altruism	Acquiescence
Analytic style	Aging quickly
Abstract thinking	Alcoholism
Artistic preference and ability	Authoritarianism
Atheism	Conservatism (of social views)
Craftwork	Crime
Creativity	Delinquency
Diet (healthy)	Dogmatism
Democratic participation (voting)	Falsification ("Lie" scores)
Educational attainment	Hysteria (versus other neuroses)
Eminence and genius	Illegitimacy
Emotional sensitivity	Impulsivity
Extra-curricular attainments	Infant mortality
Field-independence	Obesity
Height	Racial prejudice
Health, fitness, longevity	Reaction times
Humor, sense of	Religiousness
Income	Self-Esteem
Interests, depth and breadth of	Smoking

21—Dutton and Woodley of Menie, *At Our Wits' End*, *op. cit.*; developed from Arthur R. Jensen, *The g Factor: The Science of Mental Ability* (Westport, CT: Praeger, 1998).

POSITIVE CORRELATION	NEGATIVE CORRELATION
Involvement in school activities	Single/young motherhood
Leadership	Truancy
Linguistic abilities (including spelling) Logical abilities	Trust (lack of)
Marital partner, choice of	Weight/height ratio (BMI)
Media preferences	
Memory	
Migration (voluntary)	
Military rank	
Moral reasoning and development	
Motor skills	
Musical preferences and abilities	
Myopia	
Occupational status	
Occupational success	
Perceptual abilities	
Piaget-type abilities	
Practical knowledge	
Psychotherapy, response to	
Reading ability	
Social skills	
Socioeconomic status of origin Socioeconomic status achieved	
Sports participation at university Supermarket shopping ability	
Talking speed	
Trusting nature	

What are IQ Tests?

Intelligence is measured by IQ tests. We know that these are reliable because their results correlate with other intuitive measures of cognitive ability, such as educational success. Intelligence—as in the ability to solve cognitive problems quickly—increases with age, up until around middle age. In that sense even a below average 30-year-old is brighter than a very clever 3-year-old who can already read. So IQ measures your intelligence relative to others of your own age—hence, "intelligence quotient." The average person scores 100; anything less than this is below average, and anything above it is above average. Rather like height, IQ is "normally distributed" on what looks like a bell curve. Most people have an IQ of 100 with the percentages with lower or higher IQs being increasingly smaller. These are often measured in "standard deviations" of 15 points. So 68 percent of people have an IQ between 85 and 115, and 95 percent of people have an IQ between 70 and 130. This is the "normal" range. If you score below it, you are classified as retarded. If you score above it—as the kind of people reading this book probably do—then you are exceptionally bright.

The validity and reliability of IQ tests have been subject to criticism and a great deal of contemptuous dismissal from the chattering classes. American psychologist Howard Gardner has brought into question the very concept of intelligence, proposing that there are different kinds of intelligence, such as emotional, musical, and interpersonal.[22] These abilities, however, do

22—Howard Gardner, *Frames of Mind: The Theory of Multiple Intelligences* (New York: Basic Books, 1983).

not correlate with intelligence as typically defined. The exception is emotional intelligence. The ability to solve social problems has been shown to weakly positively correlate with IQ at about 0.3.[23] In other words, the more intelligent you are, the more empathetic you are; the better you can imagine what it's like to walk in someone else's shoes. So, Gardner's assorted kinds of "intelligence" are either a misuse of the word "intelligence" or they are simply examples of "intelligence" as normally defined.

Criticisms have also been leveled against IQ tests, too, but ultimately none of them really work, despite the great profundity with which they are often leveled. IQ tests have been found to have high predictive validity for school achievement (and, thus, other measures of cognitive ability), occupational status and criminality (negatively). They cannot be argued to be substantially culturally biased, as they correlate with objective measures, such as reaction times (negatively) and cranial capacity (positively): they correlate with how quick you are to respond to stimuli and simply how big your brain is, which should be unsurprising, since the brain is a thinking muscle. The robust negative correlation with reaction times (see Table 1)—how quickly you react to a stimulus, such as a light being switched on; the cleverer you are, the shorter your reaction time—implies that intelligence can essentially be reduced to a high-functioning nervous system.[24] Finally, no consistent evidence has been found for the fashionable concept of "stereotype threat." According to this theory, people who

23—Scott Barry Kaufman, Colin DeYoung, Deidre Reiss, and Jeremy Gray, "General Intelligence Predicts Reasoning Ability for Evolutionarily Familiar Content," *Intelligence* 39 (2011): 311-322.
24—Jensen, *The g Factor, op. cit.*

belong to a group believed to do badly on IQ tests—such as Black people—do badly because of their own expectation that they will do badly, which presumably stresses them out, reducing their performance. Meta-analyses have found that the evidence for this model is highly questionable. Some studies have found that groups told they will do badly on IQ tests because of some factor about them actually do *better* than expected. There is a big problem of publication bias, with studies disproving "stereotype threat" simply not being published.[25] The theory also raises the question of how exactly systematically incorrect stereotypes could originate, as 75 percent of racial stereotypes have been found to be at least partly accurate and 50 percent completely accurate.[26] Why, for instance, would Africans be stereotyped as slow witted if lived experience demonstrated the contrary? The simplest explanation, as has been empirically explored in detail by American psychologist Lee Jussim (2012), is that stereotypes develop because they are broadly true.

Are there Genetic Race Differences in Intelligence?

"Race" is one of the most controversial and taboo subjects in contemporary society, yet the concept is quite simple. A *race* is a breeding population separated from other such

25—Colleen M. Ganley, Leigh Mingle, Allison Ryan, et al. "An Examination of Stereotype Threat Effects on Girls' Mathematics Performance," *Developmental Psychology* 49 (2013), doi: 10.1037/a0031412.

26—William B. Helmreich, *The Things They Say Behind Your Back: Stereotypes and the Myths Behind Them* (New York: Doubleday, 1982).

populations for sufficient time to develop distinct sets of gene frequencies, tending to express themselves in physical and mental differences that are adaptations to their different environments. Geneticists have highlighted around 10 highly distinct genetic clusters, which are, in effect, different "races."[27] Precisely because they differ in gene frequencies, the concept of "race" allows important predictions to be made about modal intelligence, personality, the frequency of certain genetic and partly genetic medical conditions, response to drugs and drug dosage, blood type and much else. The argument that race differences are minuscule and that there are "more differences within races than between them" makes absolutely no sense. There would be tiny genetic differences between a standard musician and a Mozart, but these tiny differences, when in the same direction, come together to snowball into huge consequences. There is only a small genetic difference between chimpanzees and humans, but separating them into different species allows successful predictions to be made—this being the whole point of scientific categorization. Further, if a number of small differences all push in the same direction—because they are adaptations to a specific ecology—then this will result in clear and predictable racial differences, so it is irrelevant that there may otherwise be a high level of diversity within races.[28]

At the individual level, a meta-analysis of twin studies by Richard Lynn, whom we met earlier, found that intelligence had a heritability, on adult samples,

27—Lynn, *Race Differences in Intelligence.*
28—Gregory Cochran and Henry Harpending, *The 10,000 Year Explosion: How Civilization Accelerated Human Evolution* (New York: Basic Books, 2009).

of 0.83.[29] It might be argued that intelligence is highly heritable at the individual level but that environmental factors explain, for example, why U.S. Blacks have a lower average IQ than U.S. whites. This seems most unlikely, however, and the reasons for this have been set out by the American philosopher Michael Levin in his book *Why Race Matters*: The 15-point difference between White and Black IQ scores in the U.S. is evident by the age of 3.[30] As we've already noted, the earlier a difference becomes evident, it is argued, the more likely it is to be genetic.[31] Interracial adoption studies have shown that black adopted children's adult IQ has no relation to their white adoptive parents' IQ, but it is very similar to that of their biological parents. It is 89, whereas the Black average in the U.S. is 85.[32] The more resistant a difference is to interventions the more likely it is to be genetic. British-Canadian psychologist J. Philippe Rushton (1943-2012) noted that for millennia, outside observers, even Moorish explorers, thought that Black people had low average intelligence; and attempts to boost their intelligence, based on environmentalist assumptions, have had no significant impact.[33] Many studies have shown that race differences in IQ are on

---

29—Lynn, *Dysgenics*, 101.

30—Michael Levin, *Why Race Matters: Race Differences and What They Mean* (Oakland, VA: New Century Foundation, 2005).

31—Sarah Broman, N. Brody, P. Nichols, et al., *Retardation in Young Children* (Hillsdale, NJ: Erlbaum, 1987).

32—Richard A. Weinberg, Sandra Scarr, Irwin Waldman, "The Minnesota Trans-Racial Adoption Study: A Follow-up of IQ Test Performance at Adolescence," *Intelligence* 16 (1992): 117-135.

33—J.Philippe Rushton, *Race, Evolution and Behavior: A Life History Perspective* (New Brunswick, NJ: Transaction Publishing, 1995).

g, and thus on the most genetic parts of the IQ test, as general intelligence is strongly heritable.[34] And most importantly, moving beyond Levin's discussion, Italian anthropologist Davide Piffer has looked at the correlation between average national IQ and the average frequency in the population of genetic variants that are correlated with extremely high educational attainment, something that is very strongly associated with high IQ.[35] The correlation was 0.9, which is exceptionally high. In another study, Piffer replicated this finding with a sample of 1.1 million people from 52 countries.[36] In effect, Piffer provides extremely persuasive evidence that race differences in intelligence are overwhelmingly a reflection of genetic differences.

Critics are left with one final argument, which is that the hypothesis fits the evidence but it *must not* be true because it is "dangerous." It can be responded that this is a fallacy: an appeal to consequences and emotion. That it may be "dangerous" is irrelevant to whether or not it is true. Moreover, there are numerous serious potential dangers to building public policy around inaccurate information.

So, there are genetic race differences in IQ. This should be no more offensive to any non-White person

34—J. Philippe Rushton, "The 'Jensen Effect' and the 'Spearman-Jensen Hypothesis' of Black-White IQ Differences," *Intelligence* 26 (1998): 217-225.
35—Davide Piffer, "Evidence for Recent Polygenic Selection on Educational Attainment Inferred from GWAS Hits," *Preprints* (2016), doi:10.20944/preprints201611.0047.v1.
36—Davide Piffer, "Correlation between PGS and environmental variables," *RPubs* (2018), https://rpubs.com/Daxide/377423 (accessed August 15).

than the degree of offence I might feel if an Ashkenazi Jew asserted that Ashkenazi Jews have a higher IQ than White people. I don't feel any offence. It's a fact. Ashkenazi Jews have an average IQ of about 112.[37] In much the same way, it's a fact that university graduates have a higher average IQ than non-graduates. I can testify, however, from having attended Durham University in England and then Aberdeen University in Scotland and especially from teaching in the Cultural Anthropology Department at Oulu University in Finland, that you can get some very stupid university graduates and some very clever non-graduates. You are unlikely to find a mentally retarded university graduate or, these days at least when university attendance is widespread, a mathematical genius who has never gone to university. What is true of the group average has only a limited bearing on what a member of the group will be like.

Is There Such a Thing as "Group Selection"?

Having explored the issue of intelligence, the second concept we need to get our heads around is that of "group selection," especially because—for some reason that nobody seems to be able to coherently explain—it is "controversial" among biologists. Individual and sexual selection are well established in the field of evolutionary biology. There is individual selection—when individuals

37—Gregory Cochran, Jason Hardy, and Henry Harpending, "Natural History of Ashkenazi Intelligence," *Journal of Biosocial Science* 38 (2006): 659–693.

with certain traits are more likely to survive to pass on their genes—and there is sexual selection, when individuals with certain traits, usually indicating health, are more sexually attractive to the opposite sex, and so better able to pass on their genes. An example of sexual selection can be seen in the way that male animals will fight over a female, and the female will prefer to have sex with the winner of the fight. The winner has better genes, which will produce healthier offspring, as evidenced in the fact that he is stronger, more aggressive, and maybe even more tactical and thus more intelligent.

There is also "kin selection." People indirectly pass on their genes by aiding their kin, such as cousins, as they share 12.5 percent more of their genes with cousins than they do with random members of their ethnic group.[38] American biologists David S. Wilson and Elliot Sober have drawn upon this to advocate what they call the "Multi-Level Selection Theory."[39] They argue that once cooperative groups develop within a species, then selection will act to promote those groups that possess the optimum level of certain qualities that permit them to out-compete other groups. Thus, selection will still operate on individuals within a group but can also be seen to operate on groups themselves, as collections of individuals, and, in some circumstances, can shift away from individual and towards group selection. This model helps to explain,

38—Frank Salter, *On Genetic Interests: Family, Ethnicity and Humanity in an Age of Mass Migration* (New Brunswick, NJ: Transaction Publishers, 2006).
39—David S. Wilson and Elliot Sober, "Reintroducing Group Selection to the Human Behavioural Sciences," *Behavioral and Brain Sciences* 17 (1994): 585–654.

for example, the development of altruistic tendencies: you behave altruistically towards people, in general, because it is a way of indirectly passing on your genes; hence people tend to be increasingly altruistic to others the more closely related they are to them.[40] Kin selection involves making sacrifices for your kin and group selection is a logical extension of this, as ethnic groups are extended genetic kinship groups (Salter, 2006).

Groups tend to be selected for, under conditions of intergroup conflict, if they are internally cooperative but externally hostile. This has been demonstrated through computer modeling. Computer scientists set up a grid on which agents of four colors moved around and thus came into contact. They had a set life expectancy, but they would also die when they lost sufficient energy. They procreated asexually. When they came into contact they were programmed either to cooperate (which damaged themselves but aided the other agent) or defect (which aided themselves but damaged the other agent). The computer scientists programmed them with four options: Humanitarian (cooperate with all agents), Selfish (defect with all agents), Traitor (cooperate only with other colors), and Ethnocentric (cooperate with your own color but defect with other colors). They found that, no matter how many times they ran and tweaked the experiment, the grid eventually always came to be dominated by those operating using an ethnocentric strategy. Non-

40—See J. Philippe Rushton, "Ethnic Nationalism, Evolutionary Psychology and Genetic Similarity Theory," *Nations and Nationalism* 11 (2005): 489-507.

ethnocentric groups or individuals simply died out.[41] In addition, the more genetically endogamous they are, the more successful in-group selection groups tend to be, as the enemy group is thus more genetically different from them, and members of their own group are more similar to them, providing a stronger reason for a group member to, for example, lay down his life to protect his group. In doing so, he will help to pass on indirectly more of his genes than he would if he laid down his life for a less endogamous group.

Group selection has been criticized in depth by the well-known Canadian psychologist Steven Pinker.[42] His key criticisms are that, first, group selection deviates from the "random mutation" model inherent in Darwinian evolution. Secondly, group selection fails because, as individuals, we are clearly not going to be selected to damage our individual interests, as the theory implies. Finally, human altruistic behavior is self-interested and does not involve the kind of self-sacrifice engaged in by sterile bees. Each of these points can be answered. Firstly, if the group selection model is building on the individual selection model then it is bound to present a slightly different metaphor. To dismiss it on these grounds seems to betoken a fervent attachment to the original idea. Secondly, the group selection model merely suggests that a group will be more successful if there is genetic diversity,

---

41—Robert Axelrod and Ross A. Hammond, "The Evolution of Ethnocentric Behaviour," *Journal of Conflict Resolution* 50 (2003): 1-11, http://www-personal.umich.edu/~axe/research/AxHamm_Ethno.pdf (accessed August 15, 2019).

42—Steven Pinker, "The False Allure of Group Selection," *The Edge*, June 18, 2012, https://www.edge.org/conversation/the-false-allure-of-group-selection (accessed August 15, 2019).

meaning that an optimum percentage of its members are inclined to sacrifice themselves for their group. Thirdly, it is clearly the case that a small percentage, in many groups, is indeed prepared to sacrifice itself for the group. So, it seems to me that it is reasonable to accept multi-level selection. It's also worth noting that Pinker entitles his critique "The *False Allure* of Group Selection," as if those who regard the concept as reasonable have been somehow beguiled and bewitched, the word "allure" often having sexual connotations. In doing so, he's committing the fallacy of "poisoning the well." He's emotionally manipulating his reader to reject the concept and he's implying that those who accept it are kind of hypnotized by something seductive, a sort of fallacious appeal to insult. Perhaps Pinker himself has some deep, emotional reason for opposing the theory?[43]

So, with these important concepts clear in our minds, we can begin to explore our key question. Winston Churchill's summary implies that there is something about the nature of Islam which makes you less intelligent. Is he right?

43—Group Selection has also been criticized by physicist Gregory Cochran, who claims to have mathematically calculated that it is vanishingly unlikely to ever manifest. However, Michael Woodley of Menie has responded that, given the millions of different species on Earth, it would have come about eventually; if it is the simplest explanation for available data on how humans behave, then it very probably has done (Michael A. Woodley of Menie, "Dr. Peterson ... We Have A Problem ...," *The Jolly Heretic*, Episode 23, https://www.youtube.com/watch?v=hKQSz_2qves (accessed August 15, 2019).

Chapter 2

Religiousness and Intelligence

In their hearts is disease, so Allah has increased their disease.
—Surah 2: 10.

Religion and Intelligence

There is a very large body of evidence that religiousness is negatively associated, though weakly, with IQ in modern societies, both in terms of religious practice and the extent of religious belief. In 2014, I published the book *Religion and Intelligence: An Evolutionary Analysis*[1] in which I brought together the huge number of analyses that have been conducted on this topic since as far back as the 1920s. The evidence is overwhelming. Overall, religiousness seems to correlate with IQ at about -0.2. In other words, there is a small, though statistically significant negative association. This appears to be driven,

1—Edward Dutton, *Religion and Intelligence: An Evolutionary Analysis* (London: Ulster Institute for Social Research, 2014).

in part, by the fact that very high-IQ people tend to be particularly low in religiousness. For example, whereas, in 1997, 68 percent of British people claimed to believe in God, this was true of only 3.3 percent of members of the Royal Society, top scientists whom we would expect to be among the most intelligent people in the country. This negative association has been shown across the world, in both religious and non-religious countries, in all age groups, and even in a historical context, such as in Ancient Greece or Rome, using intelligence proxies such as education level.

It may be, however, that this relationship is not driven by general intelligence. Indeed, one study has found that the relationship between religiousness and intelligence is not actually on *g*.[2] Instead, it would appear to be due to a very narrow intelligence specialized ability which weakly correlates with general intelligence. As mentioned above, there are different kinds of intelligence, such as linguistic, spatial, and mathematical. As ability in one strongly correlates with ability in another, we are able to posit a "general factor," known as "general intelligence," via which they are related. Mathematical intelligence is a better measure of *g* than linguistic intelligence. Hence, undergraduates who study science and Mathematics tend to have higher IQs than those who pursue the Humanities, with the

---

2—Edward Dutton, Jan te Nijenhuis, Guy Madison, Dimitri van der Linden, and Daniel Metzen, "The Myth of the Stupid Believer: The Negative Religiousness-IQ Nexus is Not on General Intelligence (*g*)," *Journal of Religion and Health* (2019), doi.org/10.1007/s10943-019-00926-3.

exception of Philosophy.[3] But there is a definite *g* factor which underpins these three kinds of intelligence. Each of the three kinds of intelligence we've highlighted are composed of yet narrower "specialized abilities," whose correlation with *g* is even weaker. One of these specialized abilities is the ability to think in a highly analytical way; to look for patterns and systems. Those who are high-functioning autistics, otherwise known as sufferers from Asperger's Syndrome, are low in empathy but very high in systematizing. They see the world in a highly analytical and unemotional way, and have difficulty reading cues as to what is happening in the minds of others, even if they care about such things.[4] A review of the research into this area has shown that high-functioning autistics tend strongly towards atheism.[5]

So it may be that this is what is driving the negative relationship between religiousness and intelligence. Interestingly, it has been argued that this aspect of autism sits at one end of a spectrum, with a key aspect of schizophrenia sitting at the other. Schizophrenia is characterized, in part, by a heightened awareness of external cues of what people think: hyper-empathy. This leads to cues of friendship being interpreted as cues of profound love or cues of irritation

---

3—Dutton and Woodley of Menie, *At Our Wits' End*, *op. cit.*

4—Simon Baron-Cohen, "The Extreme Male Brain Theory of Autism," *Trends in Cognitive Sciences* 6 (2002): 248-254.

5—Edward Dutton, Salaheldin Farah Attallah Bakhiet, Khaled Elsayed Ziada, Yossry Ahmed Sayed Essa, Hamada Ali Abdelmuti Ali, Shehana Mohammed Alqafari, "Regional Differences in Intelligence in Egypt: A Country where Upper is Lower," *Journal of Biosocial Science* (2018): doi:10.1017/S0021932018000135.

being interpreted as murderous intent.[6] Schizophrenics tend to be *extremely* religious,[7] seemingly because they are obsessed with cues of the emotions of others and perceive them even in the world itself, meaning that they are prone to conspiracy theories, belief in the paranormal and intense and unusual religious belief. Those with autism spectrum disorders (ASD) simply cannot perceive any evidence of a mind behind the workings of the world and thus tend not to believe in God.

So, we can reasonably argue that a deficit in intelligence, or, at least, in a certain kind of intelligence, is associated with people being religious. In line with this, a study by myself and Dutch psychologist Dimitri Van der Linden[8] has argued that religiousness should be understood as a kind of evolved instinct or evolved cognitive bias. Religiousness is significantly genetic (around 40 percent so), it becomes more pronounced at times of stress, when people tend to become instinctive in their behavior and more prone to their cognitive biases; it exists in all cultures; it is positively associated with health and fertility; and religious experiences are associated with activity in specific brain areas. It follows that religiousness is an evolved cognitive bias. So, in other words, we have

6—Christopher Badcock, "Mentalism and Mechanism: Twin Modes of Human Cognition," In *Human Nature and Social Values: Implications of Evolutionary Psychology for Public Policy*, eds. Charles Crawford and Catherine Salman (Mahwah, NJ: Erlbaum, 2004).
7—Harold G. Koenig, "Religion, Spirituality, and Health: The Research and Clinical Implications," *ISRN Psychiatry* (2012), accessed August 15, 2019, http://dx.doi.org/10.5402/2012/278730.
8—Edward Dutton and Dimitri Van der Linden, "Why is Intelligence Negatively Associated with Religiousness?" *Evolutionary Psychological Science 3* (2017): 392-403.

been selected to be religious because it helps us to survive, for reasons we will explore later. An aspect of intelligence, we argue, is the ability to think entirely rationally and thus overcome cognitive bias or instinct. Indeed, those who were more attracted to possibilities which we are not evolved to be attracted to would be better able to solve problems, this being the essence of intelligence. This is because they would be more open to unusual possibilities; hence the correlation of 0.3 between intelligence and the personality trait termed Openness-Intellect, which involves being open-minded and intellectually curious. And these novel possibilities might solve the novel problem in question. If the problem were building a water-proof roof, then such a person might test a material that it wouldn't occur to a less open-minded person to test and that material might actually be the best material, thus better solving the problem. It follows that intelligence should predict being open to non-instinctive possibilities, modes of behavior, or thinking that we would be unlikely to have been evolved towards. Myself and Dimitri Van der Linden duly demonstrated that this is, indeed, the case. IQ predicts not just atheism, but nocturnalism, helping genetically unrelated others, and even not wanting to have any children. In other words, the higher your IQ, then the less potent will be your bias towards ethnocentric thought and behavior.

What Does This All Mean?

So, it would seem that, at the extremes, at least, religious people are highly instinctive and emotional whereas intelligent people are highly analytical. This would partly help to explain the weak negative relationship between the two concepts. It is unlikely that the relationship is simply down to "arguments for the existence of God not being logical" because, if it were, then why would intelligence also be associated with being low in other evolved cognitive biases? One also might question just how illogical certain arguments for the existence of some kind of deity, not necessarily the Christian one, are. Perhaps we are one step away from the Big Bang; the creation of an organism that itself naturally evolved. Likewise, it might be argued that religion tends to be central to civilization and once civilizations lose their religiousness, they tend to be invaded by religious foreigners, with the societies then collapsing into anarchy. This is precisely what happened with Ancient Rome and even in the Medieval Islamic world.[9] Therefore, if you value civilization, you should persuade yourself, through effortful control, to believe in God, just as you have managed to persuade yourself, despite knowing that you're just a bundle of atoms, that you "love" others and that they "love" you.

The crucial point, however, is that religiousness, in part, reflects low IQ and this is true even among religious groups. The more religious, or fundamentalist, a religious group is, the lower the average IQ of its adherents. Liberal religious groups in the U.S., such as members of Episcopal churches, have higher average IQ than fundamentalist religious groups, such as Pentecostals. Specifically, Episcopalians have an average IQ of 113,

9—Dutton and Woodley of Menie, *At Our Wits' End, op. cit.*

while Pentecostals have an average IQ of 101.[10] Consistent with the negative correlation between intelligence and conservatism, in South Korea, the least educated are Buddhist and follow the country's folk religion; Korea's oldest religion. The most educated South Koreans are Protestant (Korea's newest "world religion"), and Korean Catholics are in between. In the Netherlands, Protestants have higher average IQ than Catholics, agnostics have a higher IQ than Protestants; and atheists have the highest IQ of all.[11]

So, Do Muslims Just Reflect Low Intelligence?

This being so, could it simply be that Islam is a reflection of low intelligence? To a certain extent, this is true. Due to a complex set of factors, Islam never made it into northern Europe where (apart from in Northeast Asia) intelligence has been shown to be the highest. Lynn argues that intelligence is high in northern Europe due to its cold winters.[12] These necessitate future orientation. You must be able to conceive of a freezing day during the height of summer and plan accordingly; you need to build effective shelters and clothes; and you require the ability to create strongly bonded and cooperative groups, as these

10—Edward Dutton, *Religion and Intelligence, op. cit.*, 174-175; citing Helmuth Nyborg, "The Intelligence-Religiosity Nexus: A Representative Study of White Adolescent Americans," *Intelligence,* 37 (2009): 81-93.

11—Dutton and Van der Linden, "Why is Intelligence Negatively Associated with Religiousness?," *op. cit.*

12—Richard Lynn, *Race Differences in Intelligence, op cit.*

are more likely to survive the battle of group selection. All of these factors select in favor of high intelligence. Future-orientation is a correlate of intelligence, because people who have higher IQ are better at planning and delaying gratification. Intelligence predicts the ability to solve increasingly complex problems, such as protecting yourself from the winter cold with the best possible clothes and shelters. Strongly-bonded groups are more likely to survive the battle of group selection, as computer models have demonstrated. Intelligent people tend to be more cooperative and altruistic. These factors all help to explain why average intelligence tends to be lower among people evolved to live in hot countries, where basic needs are met to a much greater extent. Accordingly, there is an extent to which the low average IQ of Muslims simply reflects their coming from low-IQ societies. But there must be more to it, because Indians, Bangladeshis and Pakistanis are evolved to very similar conditions. Average IQ in Bangladesh is 74, in India it is 76, and in Pakistan it is 80.[13] But there's an anomaly. Southern India is hotter than northern India; basic needs are met to a greater extent. It should have a markedly lower average IQ than northern India. But it doesn't. In southern India, almost everyone is Hindu; India's 20 percent Muslim population being overwhelmingly in the north. Average IQ in southern India is 90,[14] higher than that of Greece, which is 88.[15] Indeed, the higher the Muslim percentage is in an Indian state, the lower is its average IQ. Accordingly, Cold

---

13—Lynn and Becker, *The Intelligence of Nations, op. cit.*
14—Richard Lynn and Prateek Yadav, "Differences in Cognitive Ability, Per Capita Income, Infant Mortality, Fertility and Latitude Across the Indian States," *Intelligence* 49 (2015): 179-185.
15—Lynn and Becker, *The Intelligence of Nations, op. cit.*

Winters Theory is not able to entirely explain the low IQ of Muslims. Following Cold Winters Theory, the Muslim areas of India, some of which are extremely mountainous and subject to intense cold, such as Kashmir, should have the highest IQs in India. But, in reality, they generally have among the lowest. And this is because Islam seems to negatively impact intelligence.

Muslim IQ Decreases with Age in Comparison to European IQ

Some colleagues and I[16] have demonstrated what we call the "Simber Effect." The simber is a Sudanese black stork whose arrival indicates the start of the rainy season, at the beginning of which the land is highly fertile. But this fertility does not last and, soon, the dry season hits and the land becomes arid. This metaphor appears to apply to intelligence in Muslim countries.

We obtained IQ scores from children between the ages of around 6 and 18, and in some cases all the way up to the age of 30, from representative samples in Saudi Arabia, Yemen, Egypt, Palestine, Kuwait, Libya, United Arab Emirates, Sudan, Jordan, and Qatar. We found a

16—Salaheldin Farah Attallah Bakhiet, Edward Dutton, Khalil Yousif Ali Ashaer, Yossry Ahmed Sayed Essa, Tahani Abdulrahman Muhammad Blahmar, Sultan Mohammed Hakami, and Guy Madison, "Understanding the Simber Effect: Why is the Age-Dependent Increase in Children's Cognitive Ability Smaller in Arab Countries than in Britain?," *Personality and Individual Differences* 122 (2018): 38-42.

strikingly similar pattern. Intelligence increases with age, as we have noted, but when comparing Arab and European children something very interesting happens. If we put the European IQ at 100 then the IQ of 6 year-old Arabs is about 92, placing them on a par with Greece and other relatively low-IQ countries in southeast Europe, such as Romania. However, by the age of 18, after years and years of schooling, the Arab IQ has collapsed down to just 75 points, more commensurate with Sub-Saharan African countries. By the age of 30, however, we find that Arab IQ has settled at roughly 82 points, still almost 20 points lower than Europe's.

Intriguingly, as we note in our study, you find a very similar phenomenon within the UK when you compare children whose parents are working class with children whose parents are university educated. At the age of 5, the IQ difference is small but by the age of 18, it has become a chasm. And the underlying reason is likely similar in both cases. At the age of 5 or 6, there has been relatively little time for environmental effects to hit in, so Western and Arab children's IQ difference is mainly, though not entirely, down to genes. Cold Winters Theory thus explains the slightly lower IQ of Arab children. But then these children begin formal education. In Western schools, this is oriented towards scientific and analytic thinking, precisely the kind of thinking that results in a high IQ score. Western children do not simply learn history by heart; they are taught, within certain boundaries, at least, to *think*: "Why did the English lose the Battle of Hastings? What were the key factors behind the rise of Hitler?" Accordingly, the Western education system, and, to a certain extent, simply the broader society, allows pupils to reach the phenotypic limit of their IQ, just as high nutrition allows people to reach the phenotypic limit of their height. The education system in Arab countries does not do this, nor does the broader society. Rather than stimulating children to think in a highly analytical way, it inculcates them with religious dogmas and facts that they are expected to learn by rote. As such, it does not push their IQs to their phenotypic limit.

Once people leave school, they will start to create their own environment, which reflects their own innate IQ, rather than that of school teachers and those with whom they associate at school. In some cases, this will increase their IQ, but in that average teacher IQ is about a standard deviation above the national average,[17] in most cases Arabs' IQ will start to fall and be pushed to its phenotypic maximum to a lesser extent than was the case at school. As a result, by the age of about 30, we find that Arab IQ, when compared to a British IQ of 100, is about 80 points. The social class difference in IQ in Western countries is partly explicable in a similar way. The boy from a more educated family will have a more stimulating intellectual home life. He will be talked to by his parents more and exposed to a larger vocabulary, with words being thinking tools.[18] In July 2008, while on holiday, I was sitting on a bus in Sicily one evening when I overheard an English family, with educated-sounding accents, speaking, so I listened in. The son, who was probably about 9, was trying to explain to his father why he believed his school football team, in which he played,

17—Herrnstein and Murray, *The Bell Curve, op cit.*
18—Adam Perkins, *The Welfare Trait: How State Benefits Affect Personality* (London: Palgrave Macmillan, 2015).

had lost a particular match. His argument didn't make a great deal of sense. It was very poorly expressed. However, rather than humor the boy, the father said, "I really think you need to distill what you're trying to explain down to its essence." The boy asked what "distill" and "essence" meant. In learning these terms, he could now think and communicate with slightly greater precision and subtlety than before. And, presumably, he was having conversations of this kind with his father every day. This will create an intellectual snowball effect with the IQ difference between the working-class boy and the middle-class boy getting larger and larger as they get older.

There are two caveats. Sometimes, just by genetic chance, highly intelligent people can have children who are much less intelligent than them. When this happens, such a child's IQ will be pushed to its phenotypic limit by his parents, as it will reflect the intellectual environment which they create. However, when the child becomes independent of the parental environment, and starts to create his own environment reflecting his own innate intelligence, then his IQ will fall. There will be nobody there to compel him to read books or learn to play the clarinet. There will be nobody there to arrange stimulating trips to local museums, or encourage him to "distill" his argument "down to its essence." Instead, he will find people who share his own interests, now that he is free to discover and pursue his own interests, and these will usually be people from a lower socioeconomic background than his own. Interracial adoption studies in the U.S.[19] have found that Black children will tend to be

19—Weinberg, Scarr, and Waldman, "The Minnesota Trans-Racial Adoption Study," *op cit.*, 117-135.

adopted by highly educated White couples with IQs of about 115 or higher, which pushes the Black child's IQ to its phenotypic maximum, rendering it of at least average Western IQ, if not higher. But by the time this child is an adult, its IQ tends to have fallen to about 89, not much higher than the IQ of its biological parents. This slightly higher IQ than the African American average reflects the greater intellectual stimulation the adopted child will have received during key growth phases, and, perhaps, its inculcation with certain high IQ ways of thinking, analytic thinking, by its adoptive parents.

When I was at primary school ("elementary school" in American English), I had a good friend whose parents were both university graduates. Indeed, I remember his father telling me that he was at university with the son of the former headmaster of Eton, Anthony Chenevix-Trench (1919-1979). My friend was extremely well-spoken, indeed posh, and superbly good at creative writing. But he was, tellingly, very bad at Maths; Maths being an excellent measure of *g*. He was also evidently less intelligent than his older brother whom he referred to as a "whiz kid."

We lost touch when we went to different secondary schools, but we got back in touch later, because a good school friend of mine was on the same course as him at university. In England, at that time, you studied for three A-Levels between the ages of 16 and 18 and were offered university places on the strength of your predicted A-Level grades and, in some cases, your performance in an interview at your desired university. My secondary school friend had messed up his Maths A-Level meaning that he could not take up his so-called "offers" from Oxford or the London School of Economics. Instead, on the basis of his results, he had got into a middle-ranking university, Leicester, in England's East Midlands, to do a Law Degree, this being a strongly competitive and difficult course even there. On this course, he met my primary school friend. However, as an undergraduate, my primary school friend was very much creating his own environment; his school teachers were gone and the highly intelligent parents were now peripheral. Despite working, so he told me, as hard as he could, he failed his second year and was thrown out of Leicester University. By this time, he'd also lost his posh accent and he'd become interested in things his parents would have been appalled by, such as "Mixed Martial Arts," "body building," and smoking cigars. When I was round his house in 1992, when we were 11, I had brought a WWF (World Wrestling Federation) video to watch, but his mother wouldn't allow it, sneering, "We don't believe in that." By the time I met him, a decade later, I'm sure watching *Wrestlemania VIII* would have been welcomed, in comparison to what he was now doing in his spare time.

The Education System in Kuwait

In these outlier cases, due to the high number of interacting genes involved, clever parents can produce a rather average child and, as is the case with geniuses, reasonably bright parents can produce a child of stratospheric intelligence. The other caveat is what is called "Life History Strategy;" which we will explore in more detail later. Some people simply age more quickly than others; they go through puberty earlier; they go through the menopause earlier; and they die younger, even if they live healthily. Precisely because they live life quickly, their brains will develop more quickly. As such, they may have very high IQs when they are about 6, due to their fast development, but their IQ may be about average when they are 16. In this regard, Black babies develop more quickly, in terms of growth mile stones and cognitive mile stones, than Northeast Asian babies. In other words, Black babies are more intelligent than Northeast Asian babies, because intelligence increases with age, and the Black babies are growing so quickly.[20] But by about the age of 5, the Northeast Asian children have overtaken them intellectually. This may partly explain our research group's findings when comparing Western and Arab IQs. Arab children may be following a faster Life History Strategy, which would be predicted by their evolution to a warmer environment, as we shall see, meaning that they are relatively intelligent at the age of 5, but they peak too soon; with the Western children then overtaking them to a greater and greater extent.

This may be an element to what is happening, but in that IQ scores have increased in Western countries across the twentieth century as our increasingly science-oriented environment has made us think in a more analytical way, a phenomenon known as the "Flynn Effect,"[21] the nature of Islamic society and education is still likely to be highly relevant to understanding what is going on. Across the twentieth century, IQ scores have increased by about 3 points per decade, but this has been on the least *g*-loaded parts of the test and specifically on specialized abilities

20—J.Philippe Rushton, *Race, Evolution and Behavior, op cit.*
21—James R. Flynn, *Are We Getting Smarter? Rising IQ in the Twenty-First Century* (Cambridge: Cambridge University Press, 2012).

related to analytic thinking. This has led to an overall IQ rise, despite decreases on the more *g*-loaded parts of the IQ test.[22] Consistent with the environmental influence on these rising scores, my Saudi team and I conducted a study of IQ scores among school children a decade apart and of various ages in Kuwait.[23] We found that during periods in which Islamist influence over the Kuwaiti education system, due to the composition of the government, was highest, then IQ scores tended to go down. During the period in which this influence was lowest, since about 2003; when liberals started to wrest back control from *Hadas* (the Kuwaiti branch of the Muslim Brotherhood, a fundamentalist Islamic organization), and the imitation of Western educational norms was highest, IQ tended to go up. It might be argued that Islam actually encourages intellectual inquiry into the true nature of the world. The Prophet Mohammad specifically exhorts Muslims to value knowledge of the world. It is written in the *Hadith* that: "Whoever is asked about some knowledge that he knows, then he conceals it, he will be bridled with a bridle of fire"[24]; "The seeking of knowledge is obligatory for every Muslim,"[25] and, "Acquire knowledge and impart it to the people."[26] But this can be easily interpreted, by the highly religious, as referring to knowledge that in no way contradicts anything in religious texts.

22—Dutton and Woodley of Menie, *At Our Wits' End, op cit.*

23—Edward Dutton et al., "A Negative Flynn Effect in Kuwait: The Same Effect as in Europe But With Seemingly Different Causes," *Personality and Individual Differences,* 114 (2017): 69-72.

24—*Hadith*, 2649.

25—*Hadith,* 74.

26—*Hadith,* 107.

It should be noted that a Simber Effect can also be found in developing countries which are not Muslim, such as among Xhosa in South Africa.[27] This would imply that the Simber Effect is not so much a reflection of Islam as it is of poor educational quality. But the key difference is that there is nothing inherent in Xhosa culture that leads to an education system, to the extent it can create one, which is not intellectually stimulating or analytically focused. If even the most ethnocentric Xhosa school teacher could be given the training and resources to teach in a Western fashion, then he probably would. When you follow Islam in its most rigorous and undiluted form, however, then things start to get a tad more complicated. The nature of Islam invites a poor education system, as we will see in the next chapter.

27—Salaheldin Farah Attallah Bakhiet, and Richard Lynn, "A Study of the Intelligence of Xhosa Children in South Africa," *Mankind Quarterly* 56 (2015): 335-339.

Chapter 3

Different Kinds of Religion and Where Islam Fits In

Allah! There is no god but He, the Living, the Self-Subsisting, Eternal.
—Surah 3: 2-4.

What is Religion?

Before moving on to look at the ways in which Islam influences intelligence and ethnocentrism, there is a key criticism that we need to dispatch. There is a common idea among the non-religious, among some researchers on religion, and among liberal religious types in the Church of England, the state Church where I come from, that all religions are pretty much the same. For celebrity atheist activists, like Richard Dawkins, all religion reduces stress and gives people solace by providing them with the comforting yet childish belief that God is ultimately watching over them and everything will be okay in the end. From Dawkins' perspective, this is a problem because religion is not empirically accurate, so he rails against it, in such books as *The God Delusion.*[1]

For scientific scholars of religion, all religion will tend to sanction evolutionarily adaptive behavior, such as slaughtering competitor groups who have a different religion, as God's will. In that sense, all religions are the same. The fact that evolutionarily adaptive behavior is God's will renders such behavior not just justifiable but particularly ferocious, because the enemy is literally in league with the Devil.[2] This is one of the reasons why religiousness has been shown to be a key predictor of group differences in positive and negative ethnocentrism.[3] Religiousness has been selected for, in part, because it elevates ethnocentric behavior, and it does this by sanctifying ethnocentric behavior as the eternal will of God. It understands the in-group as uniquely blessed by God and the out-group as the enemy of God. Groups that have not believed this, or have not believed it with sufficient fervor, have died out. A study by myself, Swedish psychologist Guy Madison and Richard Lynn[4] has

---

1—Richard Dawkins, *The God Delusion* (London: Bantam Press, 2006).
2—See Yael Sela, Todd K. Shackelford and James R. Liddle, "When Religion Makes It Worse: Religiously Motivated Violence as a Sexual Selection Weapon," in *The Attraction of Religion: A New Evolutionary Psychology of Religion*, eds. D. Jason Slone and James Van Slyke (London: Bloomsbury, 2015).
3—Edward Dutton, Guy Madison, and Richard Lynn, "Demographic, Economic, and Genetic Factors Related to National Differences in Ethnocentric Attitudes," *Personality and Individual Differences* 101 (2016): 137-143.
4—*Ibid.*

demonstrated that religiousness is a key factor in positively predicting how positively and negatively ethnocentric a group will be. Religious groups, and religious people, are more ethnocentric than the non-religious. Indeed, there is neurological evidence of the relationship between religiousness and ethnocentrism. In one study, an area of the brain called the posterior medial frontal cortex was rendered less active by transcranial magnetic stimulation. As a consequence, the subjects became both less negatively ethnocentric and less likely to believe in God.[5]

From the perspective of religious liberals, all religions are understood to be a response to a sense of mystery and ineffable profundity. They are a reaction to some kind of religious experience, of depth at the heart of life, beneath which lies God Himself or, for the most extreme liberals, Depth and Love which, they claim, we anthropomorphize into God. And all religions ultimately worship the same God, despite the many supposedly superficial differences. This God may be real, or He may be a sense that life has depth, an argument espoused by the Anglican bishop Rt. Rev. John A.T. Robinson (1919-1983) in his 1963 book *Honest to God*.[6] In the U.S., psychologist Helen Schucman (1909-1981) maintained in her 1975 book, *A Course of Miracles,* that we have an "inner voice" which is revealed to us as Jesus, meaning

5—Colin Holbrook, Keise Izuma, Choi Deblieck, Daniel Fessler, and Marco Iacoboni, "Neuromodulation of Group Prejudice and Religious Belief," *Social Cognitive and Affective Neuroscience* (2016), 11: 387-394.
6—John Robinson, *Honest to God* (London: SCM Press 1963).

that Jesus is, somehow, *in* us.[7] American author and quixotic presidential candidate Marianne Williamson has been heavily influenced by this book, which came to her attention during a difficult period in her life, marred by alcohol abuse and mental breakdown. Summarizing the book's message, Williamson proclaimed: "A conversion to Christ is not a conversion to Christianity. It is a conversion to a conviction of the heart. The Messiah is not a person but a point of view."[8] It has been argued that this kind of quasi-religious belief underpins the growing discipline of Religious Studies at universities. Religious Studies is usually taught as part of a broader degree in Theology, however, it has been argued that it may have become a form of extreme liberal Christian Theology itself,[9] in which all forms of religion are equal responses to the Divine, to what German philosopher Rudolf Otto (1869-1937) called the *mysterium tremendum et fascinans.*[10]

How Different Kinds of Religion Have Evolved

This viewpoint, that all religions are fundamentally the same, overlooks some very significant differences

---

7—Helen Schucman, *A Course In Miracles* (Novato, CA: Foundation for Inner Peace, 1975).
8—Leslie Bennetts, "Marianne's Faithful," *Vanity Fair*, June 1991, https://archive.vanityfair.com/article/share/35fbb1f9-a67e-401e-b55d-5da2e9dd54e3 (accessed August 15, 2019).
9—Timothy Fitzgerald, *The Ideology of Religious Studies* (Oxford: Oxford University Press, 2000).
10—Rudolf Otto, *The Idea of the Holy: An Inquiry into the Non-Rational Factor in the Idea of the Divine and Its Relation to the Rational*, trans. J. W. Harvey (Oxford: Oxford University Press, 1958).

between religions. There are different kinds of religious organization. Hunter-gatherers and pastoralists do not conceive of moral gods in the way that their successors do. They conceive of spirits who can be kind or unkind to you depending on whether you please or displease them. To please these gods or spirits, you follow the rules of the tribe by correctly carrying out its rituals, so there is an extent to which genuine religious belief will make you less likely to be cast out by the band and thus more likely to pass on your genes, helping to explain, in part, why it has been selected for. However, it must be stressed that there is a strong degree to which hunter-gatherer groups tend to be lawless and extremely violent. This is most obvious in the case of the Yanomamö of Venezuela, known by other tribes as "The Fierce People." These are mainly hunter-gatherers, though the Yanamamö do engage in a very limited degree of farming, maintaining small gardens. The Yanomamö are constantly splitting into separate groups over disagreements; husbands routinely beat and even mutilate their wives to persuade potential adulterers that they are not to be messed with; there is nothing close to the rule of law, disputes are settled by men smashing each other over the head with massive sticks until one of them gives in or dies; and the Yanomamö are notorious for slaughtering outsiders. The Yanomamö's "gods" live as they do, which, in that sense, sanctifies their evolutionarily adaptive behavior.[11]

It is only once we develop agriculture, and then city states, that this begins to change. Gods begin to mutate, such that they become concerned with people's personal morality, how they behave towards other people. This

11—See Napoleon Chagnon, *Yanomamö: The Fierce People* (New York: Holt, Rinehart & Winston, 1968).

makes sense, because we are now living in permanent settlements and in more crowded environments, so getting along with people becomes far more important. Those who believe in this moral god, who is watching their every move, are more likely to get on with other people, less likely to get into fights, and more likely to pass on their genes. By extension, the group which holds to such a religion will be more cooperative, and thus more ethnocentric, and so more likely to triumph in the battle of group selection. So, we become highly moral to our extended kinship group in a way that we didn't used to be.

But then we start to see a further subtle change. As we move towards city states, it is even more crowded, and we are increasingly having to deal with strangers, people who aren't even members of our extended kinship network. How can we be persuaded to cooperate with them, especially when we may never see them again, meaning there is a lower likelihood of reciprocity than were we to live in a village? What is the point of helping a stranger or holding your fire if a stranger offends you? The answer to this is that God becomes even more concerned with morality. He becomes universalist. You no longer cooperate with somebody because they are of your tribe, your extended genetic family, and shun everyone else. You cooperate with everybody who believes in or worships the same god that you do. Their belief in this god is a kind of insurance policy. If you both believe in the same inherently moral God, then you both believe that you will be punished for breaking that God's holy laws and, therefore, you can trust each other. Accordingly, you can, in safety, cooperate with a far larger number of people, rightly expecting that you will receive cooperation from strangers yourself.[12] This allows you to trade, to share ideas, to meet more people who are as clever as you are and intellectually stimulate each other, and to increase the gene pool. This is important because geniuses, who come up with highly original ideas that benefit their group, such as superior weapons, tend to happen by genetic chance, as we have already noted. Parents of ordinary high intelligence have a child with extraordinary cognitive ability due to unlikely, but possible, gene combinations.[13] This becomes more likely the larger the gene pool is, which is one of the reasons why Northeast Asians, despite having higher average IQ than Europeans, produce fewer per capita geniuses or important innovations. Evolved to an ecology of extreme harshness in which almost all mutations will be selected out, their gene pool is simply too small to throw up anything like as many per capita geniuses as the Northern European gene pool does, a point demonstrated in a study by Japanese economist Kenya Kura, Dutch psychologist Jan te Nijenhuis and myself.[14] Overall, gene-pool expansion means more geniuses, a greater competitive advantage and thus, again, domination of other groups in the battle of group selection.

12—See Ara Norenzayan and Azim Sharif, "The Origin and Evolution of Religious Pro-Sociality," *Science* 322 (2008): 58-62.

13—See Dutton and Woodley of Menie, *At Our Wits' End*, *op cit.*

14—Kura, te Nijenhuis, and Dutton, "Why Do Northeast Asians Win So Few Nobel Prizes?," *op cit.*

The Rise of Monotheism

This evolution towards the "moralization" of the religious heralds the rise of monotheism, a development we will explore within Judaism. There are now many different peoples cooperating and trading, and substantial differences in living standards between them. This leads to groups beginning to understand that they are economically inferior to other groups and beginning to resent that fact. This appears to lead to an intriguing change in the nature of Judaism, consistent with evidence that the two key environmental factors that appear to elevate religiousness are stress and a feeling of exclusion.[15]

Judaism moves from being a polytheistic religion, rather like modern Hinduism, in which you worship multiple gods, each of them under the chief God El, to being a "monolatric" religion. Monolatric religions accept the existence of multiple gods, but they only worship *one* of them: you identify with one particular god as being the god of *your* people and you only worship Him.[16] During the time in which the Hebrews were enslaved under the Pharaoh, this appears to be precisely what Moses did and demanded that his people do. His Ten Commandments stated that, "You shall worship no other god but I," implying that the Hebrews *did* believe that there were other gods. The Hebrews continuously drifted

15—Norenzayan and Sharif, "The Origin and Evolution of Religious Pro-Sociality," *op. cit.*
16—See Dutton and Madison, "Even 'Bigger Gods' Developed Amongst the Pastoralist Followers of Moses and Mohammed: Consistent With Uncertainty and Disadvantage Not Pro-sociality," *Behavioral and Brain Sciences*, 39 (2016).

away from this commandment. Even while Moses was still leading them, they were unconvinced of the benefits of only worshiping Yahweh and also of being thoroughly puritanical. This is recorded in Exodus 32: 1-6:

> *When the people saw that Moses was so long in coming down from the mountain, they gathered around Aaron and said, "Come, make us gods who will go before us. As for this fellow Moses who brought us up out of Egypt, we don't know what has happened to him."*
>
> *Aaron answered them, "Take off the gold earrings that your wives, your sons and your daughters are wearing, and bring them to me." So all the people took off their earrings and brought them to Aaron. He took what they handed him and made it into an idol cast in the shape of a calf, fashioning it with a tool. Then they said, "These are your gods, Israel, who brought you up out of Egypt."*
>
> *When Aaron saw this, he built an altar in front of the calf and announced, "Tomorrow there will be a festival to the Lord." So the next day the people rose early and sacrificed burnt offerings and presented fellowship offerings. Afterward they sat down to eat and drink and got up to indulge in revelry.*

But Moses insists that the Hebrews must only worship Yahweh. In doing so, he starts to change the nature of religion once again. This monolatry eventually develops into monotheism, where it is believed that there is only one god. It is unclear how this monolatry developed into a similarly intolerant form of monotheism. It may have been via the influence on Judaism of the Neo-Platonic belief in a single divine presence, the "Demiurge," a belief which became increasingly significant in Greek paganism from the time of Plato onwards.[17]

The French philosopher Alain de Benoist has observed that there are key differences between organizations that worship many gods and organizations that worship one.[18] De Benoist demonstrates that monotheistic religions have a number of key points in common. Firstly, unlike pagan religions, they do not conceive of god as an albeit mighty human with human failings and nor can people reach the status of the gods. God is entirely separate from Man and He is all-powerful and all-knowing, though not yet all-loving, because the god of the Hebrews can be extremely vengeful.

Secondly, monotheistic religions make a clear divide between "good" and "evil." They, the worshipers of the one true God, are good, and everybody else is evil. Thirdly, the world is not a morally neutral place, from which we ascend to another morally neutral place upon death. The world is an evil and Fallen place and humans are evil and Fallen, because they have disobeyed God. The flesh, and everything that is "worldly," is contaminated with this wickedness and the worldly must be rejected in favor of the godly. In paganism, by contrast, the world is morally neutral and when you die, you go to another world, not entirely dissimilar from this one. Gods themselves are worldly and have worldly aspirations. Fourthly, unlike in paganism, you do not negotiate with this God: you do

17—Colin Wells, "How Did God Get Started?" *Arion* 18 (2010): 2, https://www.bu.edu/arion/archive/volume-18/colin_wells_how_did_god_get-started/ (accessed August 15, 2019).
18—Alain de Benoist, *On Being a Pagan* (Atlanta, GA: Ultra, 2004).

exactly what He says. He is not to be argued with; obey Him or be damned. This can be seen in the way that Adam and Eve are punished for eating from the Tree of Knowledge, which would have allowed them to become as powerful as the gods. The serpent said to Eve, when she proclaimed that she must not even touch the Tree of Knowledge, let alone eat from it, lest she die, that, "You will not surely die … For God knows that when you eat of it your eyes will be opened and you will be like God, knowing good and evil" (Genesis 3: 4-5). Later God exclaims, "The man has now become like one of us, knowing good and evil. He must not be allowed to reach out his hand and take also from the tree of life and eat, and live forever" (Genesis 3: 22), clearly implying that there are many gods. God's need for total power can be seen when the Tower of Babel is knocked down and God forces humans to speak numerous different languages so that they can't understand each other. If He had not done this, then humans would have become as powerful as the gods, which could not be permitted. "Then the Lord said, 'If as one people speaking the same language they have begun to do this, nothing they plan to do will be impossible for them. Come, let us go down and confuse their language so they will not understand each other'" (Genesis 11: 5-7). Once more, we see the implication that there are many gods at this early point.

Fifthly, with this God, the very nature of time changes. Pagans believe in cycles of life; when one cycle ends then a new one begins. When you die in this world, you move on to another one and life continues much as before. But in monotheistic religions, time is linear. You have an inevitable destiny, which is a kind of paradise, assuming you fear the Lord, and one day you will get there, and be with God in paradise for eternity. Sixthly, the monotheistic religion reverses normal values, that are held by pagans: the last shall be first, the poor shall be rich, justice will prevail. In paganism, you value the eldest son best, but in the Old Testament, the figures whom we are supposed to revere are second sons; those who come second, such as Jacob, the younger brother of Esau. It is a religion of resentment where you cope with lowly status by rendering that status virtuous and godly, and so strongly elevating your own sense of self-worth and your sense of destiny. It is a religion which provides you with extreme certainty about destiny, which is very strongly focused on morality and following the rules, and which elevates the belief that your group, and only your group, is blessed by the one true God.

This actually provides a further selective advantage, because it means that, within certain boundaries at least, group membership is not achieved via ancestry, with "good" group members being those who dutifully carry out the required rituals to please the gods. Rather, group membership is via belief, via the acceptance of certain doctrines. Thus, in order to be a group member you must engage in act of intellectual submission: you must believe things in spite of them being, for example, logically impossible. You must, as Tertullian (c.155-c.240) put it, "believe it because it's absurd." This will be much easier for people who are highly cooperative and altruistic than it will be for those who are not. In addition, the group's certainty of its own righteousness tends to make the religion evangelical, meaning it spreads, whether via persuasion or coercion, over a large area. This will increase the gene pool of the core group, rendering genius yet more likely, so further elevating the group's chances in terms of group selection. Of course, there will be other factors at work, because a group that is too broad will experience other problems. It has been shown that genetic similarity is a crucial factor in explaining why people act in an altruistic, in other words, a positively ethnocentric, fashion. The closer someone is to you genetically the more likely you are to take risks for them or cooperate with them, because doing so helps to indirectly pass on more of your genes. But, so long as the optimum balance between a "genius" strategy and an ethnocentric strategy is maintained, we would expect the group to triumph.[19]

Islam and Syncretism

In other words, the nature of religion itself and, in particular, the nature of monotheism helps to strongly elevate ethnocentrism and, as we have seen, it is ethnocentric groups which tend to predominate in the battle of group selection. So, it is simply not accurate to claim that all religions are somehow the same. Perhaps they are all the same at a very fundamental level, but that could be said of many things. We can meaningfully conceive of different types of religion and Islam is very much the monotheistic type.

When groups adopt a new religion they tend to syncretize it with their earlier religion, such that, even

19—See Dutton, *Race Differences in Ethnocentrism*, *op cit.*

within Christianity, we can see glimpses of paganism.[20] This can be observed in Catholicism. Ancestor-worship appears to have been replaced by praying to saints from one's own ethnic group, such as to St Padre Pio (1887-1968) in Italy, or by organizing masses for the souls of one's ancestors. The worship of multiple gods has been replaced by the worship of the Virgin Mary, the Cult of the Saints, martyrs and angels, and belief in the Devil and assorted demons. Pagan Roman festivals have been replaced by Christian ones, which are celebrated on almost exactly the same dates. An obvious example is *Saturnalia,* which was celebrated at the same time as Christmas is now celebrated, with *Saturnalia* revelry culminating on December 25th.[21] This kind of syncretism, though not entirely absent, is far less noticeable in Protestant groups, because they are closer to the monotheistic archetype. Indeed, it could be argued that they are, therefore, closer to the Jewish archetype.

Similarly, there are different kinds of Islam. In his book *Islam Observed*,[22] the American anthropologist Clifford Geertz (1926-2006) compared Islamic practices in Indonesia and Morocco. He explored the way in which Islam has syncretized with pagan beliefs in both countries and how there are distinct forms of Islam, some highly puritanical and legalistic and others more mystical, especially Sufi Islam, operating in both countries. As with

20—See Eric Maroney, *Religious Syncretism* (London: SCM Press, 2006).

21—Michael Cooper, "Post-Constantinian Missions: Lessons from the Resurgence of Paganism," in *Contextualization and Syncretism: Navigating Cultural Currents*, ed. Gailyn Van Rheenen (Pasadena, Calf., William Carey Library, 2006), 187.

22—Clifford Geertz, *Islam Observed: Religious Development in Morocco and Indonesia* (Chicago: University of Chicago Press, 1971).

the concept of "race," it could be argued that "there are more differences within religions than there are between them." However, the same rejoinder can be made that we have made to this argument when applied to race: that there are key points of commonality, pushing in the same direction, which allow us to meaningfully distinguish between religions and certainly to distinguish between polytheistic and pagan ones.

Within the Church of England, for example, it could be argued that the extreme Low Church evangelicals, popularly known as "fundamentalists," are the most strongly monotheistic. They are the inheritors of the spirit of the Reformation and, for them, church membership is very much a matter of belief. In contrast, though doubtless a simplistic contrast, are the Anglo-Catholics. For the archetypal Anglo-Catholic, belief is not so vital. What matters is the maintenance of ritual and tradition. Then there are the Liberals, who are found within both camps. They tend not to regard Christian belief as particularly important, but they tend to be dogmatically left wing, meaning that they are held together by Leftist belief, glossed over with some Christian ritual.[23] Perhaps it could be argued that they reflect the syncretisation of Christianity with an ideology that has aspects of religiousness, such as fervent belief, a clear in-group and out-group, and a vague belief in destiny; that it is inevitable that one day a world of equality and justice will be achieved. Many researchers argued that such ideologies are "replacement religions"[24]

23—Kelvin Randall, *Evangelicals Etcetera: Conflict and Conviction in the Church of England's Parties* (London: Routledge, 2017).
24—Dutton, *Religion and Intelligence, op cit.*

because, to a certain extent, they serve a similar function to religion.

Conflicts and Splits Within Islam

Similarly, there are relatively clear divisions within Islam. There are three essential kinds of Islam: *Sunni*, *Shia* and *Ibadi*, the latter having broken away before the Sunni-Shia split. Ibadi Islam is dominant only in Oman.[25] The most significant cleavage, however, is between Sunni and Shia. This division is so fundamental that it is worth exploring in some detail. Sunni is by far the largest group, with Shia, making up about 13 percent of Muslims, being the majority only in Iran, Iraq, Bahrain, and Azerbaijan, as well as the largest minority in Kuwait, Yemen and Lebanon in which no Islamic groups constitutes a majority. Like the Catholic Church with its papal successors to Jesus, the Islamic community appointed a line of successors to Muhammad. However, there were disagreements over who these successors should be. To understand this, we need to make sense of how Islam developed.

The Prophet Muhammad (c.570-632) was born in Mecca, in what is now Saudi Arabia, and orphaned at an early age. Raised by the family of his uncle, Muhammad lived in a tribal society in which social contacts were strongly based on obligations to your clan. However, because he was orphaned, he was very much on the borders of this nomadic society, provided with just enough sustenance, merely out

25—Elizabeth Faier and Rebecca L. Torstrick, *Culture and Customs of the Arab Gulf States* (Westport, CT: Greenwood Press, 2009), 25.

of obligation to kin. The religion of Arabia at the time was mixed. Most people were polytheists or worshipped the god associated with their clan, though some people, including some of Muhammad's relatives, were Christians; religious diversity was tolerated in pagan societies, as we have seen.[26] Evidently influenced by Christianity, at the age of about 40, Muhammad was supposedly visited by the Angel Gabriel who told him that Allah had sent him as His final prophet, to confirm the monotheistic message of Abraham, Moses, the Old Testament prophets and Jesus. This message was one of complete "submission" ("*Islam*") to Allah. From this point onwards, and for the rest of his life, he continued to have religious visions in which Allah's instructions were given to him. These became the *Koran.* Muhammad attempted to convert the people of Mecca to Islam, but they reacted to this aggressively. So Muhammad fled, with the band of followers he had now gathered, to Medina. In Medina, Muhammad managed to unite the various warring tribes behind him and convert them to Islam. The Prophet then returned to Mecca with an army about 10,000 strong and took over the city. By the end of his life, Muhammad's armies had conquered and converted the entire Arabian Peninsula and beyond, all the way up to Damascus. When Muhammad died, a successor as leader, "Caliph," of the Islamic community was chosen and the conquests continued.[27]

It was this change of leadership that led to Islam's fundamental split. Before the Prophet Muhammad died he appointed his cousin and son-in-law, Ali (601-661)

26—Benoist, *On Being a Pagan, op. cit.*
27—See Karen Armstrong, *Muhammad: A Biography of the Prophet* (New York: HarperCollins, 1993).

as his successor as Caliph of the Islamic state. ("Caliph" effectively means "successor to Muhammad." "Emir" means "Commander." However, at various points in history, leadership has been dominated by a particular family meaning that these have effectively become hereditary positions, akin to the concept of "King"). After Muhammad died, a dissident group pledged their allegiance to Muhammad's father-in-law, Abu Bakr (c. 573-634). Bakr assumed political control, becoming the first Caliph. Those who accepted this were to become the *Sunni.* Those who did not, and pledged their allegiance to Ali, became known as the *Shia*; which translates as "followers of Ali." The Shia regarded Ali and his blood heirs via Muhammad's daughter Fatima, who were thus descendants of the Prophet Muhammad, as Islam's religious authorities, no matter who wielded political power. After Bakr and some of his successors died, there was a power vacuum and the Sunni pledged their allegiance to Ali, rendering him leader both politically and religiously. Ali was succeeded by his son Hassan ibn Ali (624-670), who was the Prophet Muhammad's grandson. But, after six months in power, Hassan ibn Ali was compelled, by military force, to make a treaty with Muhammad's brother-in-law Muawiyah I (602-680), who became the fifth Caliph. Under the terms of the treaty, Muawiyah I would have political power, but he wasn't allowed to select his own successor. He broke this treaty, nominating his son Yazid I (647-683), leading to the creation of the Umayyad Dynasty and Caliphate, which was fully accepted by Sunnis.

As far as the Shias were concerned, Yazid I was not the religious successor to Muhammad. The successor was

Hassan's brother, Husayn ibn Ali (625-680), and then his successors, who were blood descendants of the Prophet Muhammad. Ultimately, this meant that there were twelve "Hidden Imams" (meaning "Hidden Religious Leaders" rather than the way in which the word "imam" is commonly understood) beginning with Hassan ibn Ali, whom the Shia regarded as religious authorities, though the Sunni did not. These "Hidden Imams" did not have any overt political power. They were not caliphs. Accordingly, mainstream Shias became known as "The Twelvers." They became a bit like Methodists in eighteenth century England. The Shias did not accept the religious pronouncements of the bishops, who represented the state Church. They followed the teachings of a purely religious leader who was not part of the state apparatus, let alone the leader of this state apparatus.[28] This split gradually led to theological differences, because for centuries Shia and Sunni Muslims had different religious leaders. The Sunni believe that the successor to Muhammad, both religiously and politically, is the Caliphate; with its leader chosen from among the ablest men. The Shia believe that the successor to Muhammad is the Imamate. As we will see below, this is eternally headed by the *Mahdi,* the last "Hidden Imam" whom, they believe, never died and who will reveal himself at the end of time.[29] In contrast, Sunnis believe that the *Mahdi* is a man whom Allah will send to make the entire world Islamic. The follower's of Sudan's

28—Herbert Brook Workman, *Methodism* (Cambridge: Cambridge University Press, 1912).
29—Ali Abdel Razek, *Islam and the Foundations of Political Power* (Edinburgh: Edinburgh University Press, 2013).

Mahdi believed that he was this very man.[30] Also, for Shias, Allah will also periodically send so-called "Imams," who are infallible, to put Shia Islam back on the right path. These Imams will always be blood descendants of the Prophet Muhammad via Fatima and Ali.[31]

There are many differences in terminology between Sunni and Shia as well. In media reports about Islamic terrorism, Islamic forms of address, think "Mullah Omar" or "Ayatollah Khomeini," are frequently employed without any discussion of what they mean, so it would be helpful to be clear. As already touched upon, in Shia Islam, the word "Imam" has a very different meaning from in Sunni Islam. In Shia Islam, the Imam is the single religious authority among Muslims. The spiritual successor to Muhammad, the Imam is divinely inspired and infallible, and there can only be one Imam at any given point, if there is one at all. The Ayatollah Khomeini was believed by his followers to be his era's Imam. Shia Islam, rather like the Catholic Church with its hierarchy of Pope, Cardinals, Archbishops, Bishops and priests, involves a graded hierarchy of religious practitioners. However, unlike with the Catholic Church, there are many such competing hierarchies, each of them led by a *Marja*. A *Marja* is a cleric whom a portion of Shia believe to be the ultimate religious authority, to the extent that they are prepared to pay their religious taxes to him and obey his *fatwas* (religious edicts). Obviously, some *Marja,* such as the Ayatollah Khomeini (1902-1989), become particularly

30—See Haim Shaked, *The Life of the Sudanese Mahdi* (London: Routledge, 2017).

31—Sami Zubaidi, *Islam, the People and the State: Political Ideas and Movements in the Middle East* (London: I. B. Tauris, 1993), 35, note 2.

influential because their following is so substantial; so many Shias regard them as the supreme religious authority. Beneath the *Marja*, in order of precedence and perceived theological knowledge, are the *Grand Ayatollahs, Ayatollahs, Hojatoleslams, Sheikhs* (roughly comparable to an imam in Sunni Islam; see below) and *Mullahs*, who are effectively lesser clerics.[32] They achieve these ranks, in effect, by community acclamation.

In Sunni Islam, there is no such hierarchy, so it can perhaps be compared to the Calvinist churches in that sense. Apart from among the Sufi (see below), the word "imam," in Sunni Islam, simply refers to the cleric who leads the mosque in prayer and delivers sermons. A man will be appointed an imam based on his theological knowledge, education, and charisma. He is not a priest. There is no sense in which he is believed to mediate between humans and deities via rituals. So he is best compared to the pastor of a free church or to a Rabbi. In Somalia, which is mainly Sunni, the word *Sheikh*, which in much of the Sunni Arab world refers to an aristocratic tribal leader, is used instead of "imam."[33] In Sunni Islamic societies, the *Grand Mufti* is the imam or judge recognized by the state, or the Islamic community, as being the leading authority on religious matters. A *Mullah*, in Sunni Islam, is simply a person who is highly educated in theology, such as a Sharia Law judge (Sharia being the system of laws laid down by the Prophet Muhammad) or a theology teacher. Thus, a *Mullah* has a higher status in Sunni than in Shia Islam.

32—Edgar O'Ballance, *Islamic Fundamentalist Terrorism, 1979-95: The Iranian Connection* (Basingstoke: MacMillan Press, 1997), 46.
33—Ioan M. Lewis, *Saints and Somalis: Popular Islam in a Clan-based Society* (Lawrenceville, NJ: The Red Sea Press, 1998), 59.

An *Alim,* in both Sunni and Shia, is a minor religious minister and also an Islamic scholar.

As we have seen, the Shia, though *not* the Sunni, also maintain a belief that the last "Hidden Imam" is present in the world. Shia's eleventh imam, Hasan ibn Muhammad (846-874), was poisoned on the orders of the Caliph, and Hasan's son, Muhammad ibn Hasan al Madhi (868-874?), went missing. Shia believe that this "Twelfth imam" never died and will, one day, return, in some kind of reincarnation known as the Mahdi, to Islamize the world and restore rightful leadership to the Islamic community.[34] There are, in addition, numerous Shia sects. For example, 42 percent of Muslims in Yemen are *Zaydi*, this split having been led by the son of the fourth Hidden Imam.[35] The *Ismailis* derive their name from their belief that the rightful successor to the sixth Hidden Imam was Ismail Ibd Jaffar (719-762), meaning they are known as the "Seveners." Their thought is heavily influenced by Neo-Platonism, in some ways a kind of pantheism.[36] The largest Ismaeli sect are the *Nizaris*, who make up about 10 percent of Shias. They regard the hereditary *Aga Khan* as the Imam.[37]

In terms of "liberal" Shia groups, there is the *Alawites.* Dominant on the Syrian coast, members

34—Robert Brenton Betts, *The Sunni-Shi'a Divide: Islam's Internal Divisions and Their Global Consequences* (Washington, DC: Potmac Books, 2013).

35—Stephen W. Day, *Regionalism and Rebellion in Yemen: A Troubled National Union* (Cambridge: Cambridge University Press, 2012), 31.

36—Jonah Blank, *Mullahs on the Mainframe: Islam and Modernity Among the Daudi Bohras* (Chicago: University of Chicago Press, 2001), 169.

37—Innes Bowen, *Medina in Birmingham, Najaf in Brent: Inside British Islam* (Oxford: Oxford University Press, 2014), 167.

reject literal interpretations of the *Koran* and they also reject the *hijab*. Where orthodox Muslims shun alcohol, *Alawites* drink wine not only in general but also in their rituals, believing it to be the transubstantiation of Ali whom they effectively worship. In addition, they believe that males can reincarnate.[38] Within Sunni Islam, there developed *Sufism*, that has many points in common with Medieval Roman Catholicism. However, Sufism is focused around mysticism, to the extent that many Sufi believe that they are free from Islamic Law by virtue of their direct contact with Allah and Muhammad through mystical experience. They achieve this at their mosques through entering ecstatic trances, often via chanting and dancing. This, they believe, allows them to attain the correct interpretation of Islamic texts.[39]

Sufism tends to be syncretized with pre-Islamic cultural practices more than other forms of Islam, as well as being more tolerant of pluralism. The Sufi venerate saints, in the form of the first three generations of Islamic leadership, as well as various Sufi mystics associated with miracles who are regarded as saints. In this regard, it must be remembered that many parts of the Islamic world were Christian when Islam was introduced, meaning that they have syncretized Christianity and Islam. Sufis pray to saints as intercessors between themselves and God, just as many Roman Catholics do.

38—Yaron Friedman, *The Nusayri Alawis: An Introduction to the Religion, History and Identity of the Leading Minority in Syria* (Leiden: BRILL, 2010).
39—Zackery M. Heern, *The Emergence of Modern Shi'ism: Islamic Reform in Iraq and Iran* (London: Oneworld Publications, 2015), 142.

Many Sufi travel to holy places or saintly shrines where they believe that a saint or Allah can better hear their prayers; they pray at the shrines of certain venerated Sufi sheikhs (leaders); and they pray over the relics of these sheikhs.[40] The dominant form of Islam during the heights of Islamic Civilization, Sufi is based around a series of Sufi orders or brotherhoods, with initiates into these orders passing on the knowledge and practices they learn in a line of teaching that can ultimately be traced back to Muhammad.[41] The leaders of Sufi orders are known as Sheikhs, a title which was eventually extended to Sufi imams as well.[42]

There are many conflicts between these different groups. Fundamentalist Sunnis, whom we will explore below, regard Shias as heretics who worship Ali.[43] They also brand Sufi as heretics, persecuting them and sending suicide bombers to their shrines, as has occurred in Pakistan in recent times.[44] The Shia have sometimes persecuted the Sunni, such as in Persia, and most recently

40—H. T. Norris, *Popular Sufism in Eastern Europe: Sufi Brotherhood and the Dialogue with Christianity and the Heterodoxy* (London: Routledge, 2007).

41—Nile Green, *Sufism: A Global History* (Hoboken, NJ: John Wiley & Sons, 2012).

42—Akbar Ahmed, *Journey Into Europe: Islam, Immigration and Identity* (Washington, DC: Brookings Institution Press, 2012), 195; Anna Zelkina, *In Quest for God and Freedom: The Sufi Response to the Russian Advance in the North Caucasus* (London: C. Hurst & Co., 2000), 47.

43—Zackery Heern, *The Emergence of Modern Shiism: Islamic Reform in Iraq and Iran.* London: Oneworld Publications, 2015), 96.

44—Meena Sharify-Funk, Merin Shobhana Xavier, and William Rory Dickson, *Contemporary Sufism: Piety, Politics, and Popular Culture* (London: Routledge, 2017), Ch 2.

in Iraq, when the majority Shia took over after decades of minority rule by Saddam Hussein's (1937-2006) Sunni-dominated regime between 1979 and 2003.[45]

Conflicts and Splits Within the Caliphate

There were also numerous political splits within the Islamic empire. Any particular Caliphate struggled to control its territory, with parts of it being more loyal than others and parts of it, in practice, being beyond the Caliphate's control, effectively religiously and politically independent. This can perhaps be compared to the way in which, in theory, the Pope was in charge of Medieval Christendom, but the reality was rather more complicated.

The united Umayyad Caliphate was overthrown in 750 by the Abbasid Dynasty, who claimed descent from one of Muhammad's uncles. They moved the capital from Mecca to Baghdad, with the dynasty lasting until the fall of Baghdad to the Mongols in 1258. Under the theoretical religious and political authority of the Caliph in Baghdad, there were many Emirs who ruled smaller areas. Indeed, there were further localized emirs who ruled very small areas, paying fealty to the emir of the larger area in which their emirate lay. However, various emirs openly declared independence at various points, meaning there was no religious or political authority higher than

45—Anthony Celso, "How to Defeat the Islamic State: Crafting a Rational War Strategy," in *Jihadi Terrorism, Insurgency and the Islamic State: A Small Wars Journal Anthropology*, ed. Dave Dilegge and Robert Bunker, Vol. III.

them even in theory. In 912, the Emir of Cordoba, Abd ar-Rahman III (c.889-961), declared himself Caliph of Cordoba, extending his rule over almost all of Iberia. This caliphate collapsed in 1231 into numerous small Muslim kingdoms. In 909, the Fatamid Caliphate was established by a Shia in Egypt, gradually extending its rule to Maghreb (Northwest Africa), Sicily, Sudan, parts of the Levant and even to the Western portion of Saudi Arabia, which includes Mecca. This caliphate collapsed in 1171. The Almohad Caliphate ran parts of North Africa and Iberia between 1145 and 1269. And, most importantly, the Ottoman Caliphate was established in 1517 and ruled until 1924, extending its rule into the Balkans.[46]

In addition, many Messianic figures have promoted themselves as Mahdi and established a Caliphate accordingly, though most of these have fizzled out. Of the more successful ones, Syed Muḥammad Jaunpuri (1443-1505), from Uttar Pradesh in northern India, declared himself Mahdi at Mecca and has followers to this day in the *Mahdavi* sect.[47] We have already discussed the relatively short lived Caliphate in Sudan at the end of the nineteenth century. Coincidentally, Churchill was present at the Battle of Omdermun in 1899, when British forces comprehensively crushed it.[48] In 1889, Punjabi writer Mirza Ghulam Ahmad (1835-1938), who had declared himself Mahdi in 1875, established the

46—Hugh Kennedy, *The Caliphate: The History of an Idea* (London: Penguin, 2016).
47—A. Azfar Moin, *The Millennial Sovereign: Sacred Kingship and Sainthood in Islam* (New York: Columbia University Press, 2012), 159.
48—Simon Read, *Winston Churchill Reporting: Adventures of a Young War Correspondent* (Boston: De Capo Press, 2015), chapter 8.

Ahmadiyya community. By the time of his death, he had 40,000 followers and his "Caliphate" continues to this day, though it lacks direct political power.[49] Arab silk merchant Wallace D. Fard (1877?-1934?) declared himself the Mahdi and launched the *Nation of Islam* in Detroit in 1930.[50] The group is now based in Chicago, their mosque being an impressive converted Greek Orthodox church which I visited in June 2019. Very much an Islamic sect with an idiosyncratic theology, it is overwhelmingly African-American and it is not open to White people.[51] However, though cautious at first, they were eventually very friendly upon my visit, even giving me one of their delicious "bean pies."[52] In 2006, Abu Bakr al-Baghdadi (1971-2019) set himself up as Caliph of the Islamic State of Iraq and Syria (ISIS), which, at one point, held a third of Iraqi territory, though it is now very much diminished, having lost 98 percent of its territory since its peak in 2014.[53] In 2014, al-Baghdadi also declared a worldwide Caliphate with him as the world's Caliph.[54]

49—Simon Ross Valentine, *Islam and the Ahmadiyya Jama'at: History, Belief, Practice* (New York: Colombia University Press, 2008).

50—Martha F. Lee, *The Nation of Islam: An American Millenarian Movement* (Syracuse, NY: Syracuse University Press, 1996).

51—Dawn-Marie Gibson, *History of the Nation of Islam: Race, Islam, and the Quest for Freedom* (Santa Barbara, CA: Praeger, 2012).

52—Edward E. Curtis, *Black Muslim Religion in the Nation of Islam, 1960-1975* (Chapel Hill, NC: University of North Carolina Press, 2009), 107.

53—Jamie McIntyre, "Here's How Much Ground ISIS Has Lost Since Trump Took Over," *Washington Examiner,* December 23, 2017, https://www.washingtonexaminer.com/heres-how-much-ground-isis-has-lost-since-trump-took-over (accessed August 15, 2019).

54—Will McCants, *The ISIS Apocalypse: The History, Strategy, and Doomsday Vision of the Islamic State* (New York: St Martin's Press, 2015).

Also, in a number of Islamic countries, there are substantial areas that are not under the control of the internationally recognized government. Instead, these regions are run by assorted "terrorist" organizations. As far as these organizations are concerned, the territory they control is their "emirate" and their leader is the "Emir." Thus, the head of *Al Qaeda*, at the time of writing Egyptian surgeon Ayman Al-Zawahiri (b.1951), is the Emir of the areas that the group and its allies govern. He has appointed various regional emirs, who swear fealty to him, to control smaller regions within his emirate.[55] In Somalia, for example, the Emir of the Al Qaeda-allied terrorist group *Al Shabaab*, since 2014, has been Ahmed Omar (b.1972). He is thus the Emir of significant portions of Somalia,[56] though he ultimately pays fealty to Al-Zawahiri, who succeeded Osama Bin Laden (1957-2011) as Al Qaeda's Emir.

In the Sunni world, *Wahhabism*, a kind of fundamentalist revival which began in the eighteenth century and which is the state doctrine of Saudi Arabia, can be classified as extremely conservative. Indeed, Wahhabists were originally known as the "monotheists."[57] The movement was named after its leader Muhammad Al Wahhab (1703-1792). Wahhab wanted to cleanse Islam of practices such as

55—Joscelyn, "Al Qaeda is Very Much Alive and Widely Misunderstood," *The Weekly Standard*, September 11, 2018, https://www.weeklystandard.com/thomas-joscelyn/sept-11-anniversary-17-years-later-al-qaeda-is-alive (accessed August 15, 2019).

56—Hamza Mohamed, "Meeting Al-Shabab's Elusive Leader," *Al Jazeera*, January 22, 2016, https://www.aljazeera.com/blogs/africa/2016/01/meeting-al-shabab-elusive-leader-160121130900746.html (accessed August 15, 2019).

57—Heern, *The Emergence of Modern Shiism, op. cit.*, 146.

the veneration of saints (with early followers of Muhammad venerated as such) and of idolatry. In 1744, this Arab Luther, who was living under Ottoman Rule, established, in alliance with Muhammad bin Saud (d. 1765), the state that would one day become Saudi Arabia; the Emirate of Diriyah. Saud was the first Emir, and Wahhab the leading cleric. To this day, the Sauds are the political rulers and Wahhab's descendants, the House of Shaykh, are effectively in charge of religious issues. The Ottomans recaptured the territory in 1818, only for it to break away again in 1824.[58] In the nineteenth century, with Egypt under British rule, there developed a proto-fundamentalist reaction against Westernization in Egypt known as *Salafism*. The *Salafists* aimed to restore the supposedly authentic and pure Islam practiced by the *Salaf*; the first three generations of Muslims. Salafists were themselves influenced by Wahhabism, which they regarded as a form of proto-Salafism.[59]

Wahhabism and Salafism were vital influences on the Egyptian theologian who is central to contemporary "radical Islam," Sayyid Qutb (1906-1966). Qutb established the fundamentalist Muslim Brotherhood and was eventually executed by Egypt's secular government. He spent time in the U.S. and concluded that it was a decadent, self-destructive culture; one which was being exported to Muslim countries.[60] "Humanity today is living in a large brothel!" Qutb declared. In such a "rubbish

58—As'ad Abukhalil, *The Battle for Saudi Arabia: Royalty, Fundamentalism, and Global Power* (New York: Seven Stories Press, 2011).
59—Namira Nahouza, *Wahhabism and the Rise of the New Salafists: Theology, Power and Sunni Islam* (London: I.B. Tauris, 2018).
60—Sayyid Qutb, *Milestones* (Chicago: Kazi Publications, 1964).

heap," nobody was free to choose their faith because their minds were sullied by the surrounding cesspool of sin. Muslims, therefore, had a duty to "step forward and take control of the political authority."[61] Accordingly, Qutb advocated a policy of Islamism; the reordering of society along fundamentalist Islamic and Sharia lines. In order to achieve this society, he advocated violent *Jihad*; Holy War.[62] Qutb's ideas thus extended Wahhabism in a revolutionary new direction. Whereas Al-Wahhab was concerned with proper worship and belief with reference to Islamic societies, Qutb wanted the entire world to adopt this system.[63] Qutb's ideas are ultimately behind the Jihad waged by such groups as Al Qaeda and other Islamic terrorist groups worldwide.[64]

By contrast, in Near Eastern countries, especially, Sunni Islam is more ritualistic, Sufi-based, and more clearly syncretized with paganism,[65] though there is growing Wahhabi influence in all Sunni Muslim countries and communities.[66] I live in Oulu in northern Finland. In 2009, I conducted participant observation fieldwork at the city's small mosque. One evening, at my Finnish

---

61—Mark Weston, *Prophets and Princes: Saudi Arabia from Muhammad to the Present* (Hoboken, NJ: John Wiley & Sons, 2008), 366.

62—James Toth, *Sayyid Qutb: The Life and Legacy of a Radical Islamic Intellectual* (Oxford: Oxford University Press, 2013).

63—David Commins, *The Wahhabi Mission and Saudi Arabia* (London: I.B. Tauris, 2008).

64—John Calvert, *Sayyid Qutb and the Origins of Radical Islamism* (Oxford: Oxford University Press, 2009).

65—Maroney, *Religious Syncretism op. cit.*

66—Natana J. DeLong-Bas, *Wahhabi Islam: From Revival and Reform to Global Jihad* (London: I.B. Tauris, 2007).

language course, I got talking to a Tajik (Near Eastern) doctor about this mosque. He informed me that he would never attend such a mosque, because it was "Wahhabist"[67] and the imam, who was extremely welcoming as were his congregants, openly admitted to me connections to organizations related to the fundamentalist Muslim Brotherhood. Indeed, it was reported in the Bangladeshi press in 2017 that a Bangladeshi called Taz Rahman moved to Finland, "met Dr Abdul Mannan, imam of a mosque in Oulu, Finland. He became influenced by Mannan's radical religious views," married Mannan's daughter, and was killed in 2017, fighting for ISIS in Iraq.[68] This was of particular interest in Finland, at least in various conservative news forums, because Mannan was not just Oulu's imam but, at that time, a Social Democrat deputy member of its city council.[69] The Finnish mainstream media did not report my findings, nor those of the Bangladeshi press, until May 2019[70] in the wake of a Muslim child grooming scandal

67—Edward Dutton, *Four Immigrant Churches and a Mosque: An Overview of Immigrant Religious Institutions in Oulu* (Turku: Finnish Institute of Migration, 2009), 32.

68—Nuruzzaman Labu, "How the Tattooed, Party-loving Taz Became Abu Ismail al-Bengali," *Dhaka Tribune*, May 18, 2017, https://www.dhakatribune.com/bangladesh/crime/2017/05/18/tattooed-party-loving-taz-became-abu-ismail-al-bengali (accessed August 15, 2019).

69—*Mvlehti Media*, Oulussa asunut Isiksen jihadisti Taz Rahman kuollut Irakissa—miehen appi on oululainen imaami, May 12, 2017, https://mvlehti.net/2017/05/12/oulussa-asunut-isiksen-jihadisti-taz-rahman-kuollut-irakissa-miehen-appi-on-oululainen-imaami/ (accessed August 15, 2019).

70—Mikko Marttinen," Oulun Moskeijaa Rahoitetaan Saudi-Arabiasta," May 22, 2019. *Ilta-Sanomat.*

in Oulu[71] when Finland's largest tabloid, *Ilta-Sanomat,* also managed to discover that the mosque received funding from Wahhabi Saudi Arabia.

Within Shia, there is a similar fundamentalist movement. Throughout its history, various revolutionary and strongly religious figures have been presumed to be The Mahdi. In addition, the Ayatollah Khomeini (1902-1989), who turned Iran into a fundamentalist Islamic state in 1979, is regarded, in Iran, as an Imam, a rightful successor to Muhammad, who has supreme religious authority. He was also a *Marja;* the highest form of religious authority after an Imam.[72] As with Salafism in Sunni Islam, Khomeini's Shia movement was an extreme conservative reaction against the rapid pace at which Iran had Westernized and modernised; this reaction against the modern world and concomitant attempt to revive an idealized past often being the essence of religious fundamentalism. Accordingly, Khomeini's supporters overthrew the Shah of Iran, Mohammad Reza Pahlavi (1919-1980), in February 1979's "Islamic Revolution." Khomeini duly returned from exile to rule what would become a Shia Theocracy. In Iraq, the *de facto* fundamentalist Shia political leader, though he is not a *Marja,* is the cleric Muqtada al-Sadr (b.1974), who has his own militia and is extremely influential, to the extent of having his own parliamentary list, which garners

71—See Edward Dutton, *The Silent Rape Epidemic: How the Finns Were Groomed to Love Their Abusers* (Oulu: Thomas Edward Press, 2019).
72—Baqer Moin, *Khomeini: Life of the Ayatollah* (London: I.B. Tauris, 1999).

considerable electoral support.[1] Sadr is a descendant of the Prophet Muhammad.[2]

What Would Muhammad Do?

Evidently, then, there are many different kinds of Islam, just as there are many different kinds of Christianity. However, there are two key factors with regard to Islam, and they are factors that, as we will see, help to ensure that Islam assists in reducing intelligence and elevating ethnocentrism, at least if practiced in an orthodox fashion, in a way which is distinct from Christianity and other religions.

The first of these is the fact that, according to Islam, everybody is born Muslim.[3] Muhammad's message is universal and, as such, you do not "convert" to Islam; you "revert."[4] In all other religious groups, you "convert" by being initiated into a community of religious practice which is accepted as being part of the religion in question. Few people, least of all archetypal Christians, would accept that you were a Christian if you simply went around

---

1—Patrick Cockburn, *Muqtada al-Sadr and the Fall of Iraq* (London: Faber & Faber, 2015).

2—Jane Arraf, "Iraq Works to Form a New Government," May 23, 2018, *National Public Radio,* https://www.npr.org/2018/05/23/613596998/iraq-works-to-form-new-government?t=1538987153305 (accessed August 15, 2019).

3—Danial Zainal Abidin, *Islam: The Misunderstood Religion* (Selangor, Malaysia: PTS Millennia, 2007), 168.

4—Scott Flower, *Islam and Cultural Change in Papua New Guinea* (London: Routledge, 2016), 50.

saying you were. To be a Christian, in any conventional sense of the word, you would need to at least be baptized into a specific denomination. Accordingly, you would thus have to confess to believe not merely that "Jesus is a paradigm to be imitated," but the specific dogmas of that particular denomination. In Islam, this is simply not the case. You are a Muslim if you genuinely accept that there is "no other God but Allah and Muhammad is His prophet." This is a fundamental Islamic dogma. You are Muslim if you believe that Muhammad speaks for God and, by implication, you must, therefore, do what Muhammad does and accept what he says is God's will. Moreover, a second essential Islamic doctrine is that if the *Koran* has Muhammad contradict himself, then the more recent statement, the one closer to the end of the *Koran,* is the correct one, as this is Muhammad's ultimate decision, something known as the Doctrine of Abrogation. This doctrine is explicitly stated in the *Koran* itself, meaning it is regarded as emanating from the Prophet Muhammad, and thus Allah himself.[5] Importantly, however, you do not need to be initiated into a specific Islamic group, which may have developed their own syncretized or tendentious understanding of Islam.

The second of these factors is the centrality of Muhammad, combined with all that is known about him. The Ancient Greek historian Euhemerus (4th-3rd century BC) speculated that all religions could ultimately be traced back to the kind of shaman figures, with their

5—David Bukay, "Peace or Jihad? Abrogation in Islam," *Middle East Quarterly* 14 (2007): 3-11, https://www.meforum.org/1754/peace-or-jihad-abrogation-in-islam (accessed August 15, 2019).

healing powers, trances, and supposed miracles,[6] that you find in primitive social organizations. He suggested that over time, these people were mythologized into gods and that Zeus had existed historically and had been subject to precisely this process of "Euhemerism."[7] However, figures such as Zeus are not to be slavishly imitated in all respects. They are gods, but they have distinct personalities and they accordingly have human frailties. Even in Judaism, though Moses and many other Old Testament prophets are strongly revered, they are perceived as real people who do good and bad things. Moses, for example, loses his temper and murders people in a surreptitious manner. "He saw an Egyptian beating a Hebrew, one of his own people. Glancing this way and that and seeing no one, he killed the Egyptian and hid him in the sand" (Exodus 3: 11-12). Even Yahweh, as portrayed in the Old Testament, possesses human qualities which few followers would regard as positive: "Do not worship any other god, for the Lord, whose name is Jealous, is a jealous God" (Exodus 34: 14).

With Jesus we see the development of the idea that the object of worship is a paradigm of humanity. Not only is Jesus actually fully God, as part of the Trinity, but he is also fully human and he is perfect human. Accordingly, he is to be imitated in every possible way.[8] Even Christ's supposed celibacy is supposed to be imitated if at all possible, though Saint Paul emphasizes that this is very

6—Peter Vitesbsky, *Shamanism* (Norman, OK: University of Oklahoma Press, 2001).

7—Stylianos Spyridakis, "Zeus Is Dead: Euhemerus and Crete," *The Classical Journal* 63 (1968), 337–340.

8—See John Lawson, *Introduction to Christian Doctrine* (Grand Rapids, MI: Francis Ashbury Press, 1999).

difficult, so if you really have to then you should get married and have children:

> *I'm telling those who are single and widows that it's good for them to stay single like me. But if they can't control themselves, they should get married, because it's better to marry than to burn with passion.* —I Corinthians 7: 8-9

Jesus is "perfect man" and you should live your life constantly asking yourself, "What Would Jesus Do?" –WWJD; a phrase printed on wrist bands worn by evangelical Christians. But, crucially, we don't really know what Jesus *would* do, because our information about him is extremely scant, limited to a few reports about his life written between 40 and 100 years after he died, and to his own supposed parables, preserved in the four Gospels. There exist no eye-witness accounts about Jesus at all, not a single primary source remains.[9]

However, when it comes to Muhammad the situation is rather different. Muhammad is "perfect man"; to be imitated by all good Muslims.[10] Partly because he arrived 500 years later than Jesus, we have considerable detail about Muhammad's life, how Muhammad behaved, and what he thought. This has been preserved in the Islamic holy texts. We have records of Muhammad's thoughts and actions, supposedly orally transmitted from those

9—Delbert Royce Burkett, *An Introduction to the New Testament and the Origins of Christianity.* Cambridge: Cambridge University Press, 2002).

10—Daniel W. Brown, *A New Introduction to Islam* (Hoboken, NJ: John Wiley & Sons, 2011), 205.

who knew him and written down about two generations after Muhammad's time. These records are known as the *Sunnah.* The *Sunnah* is composed of the *Hadith* and the *Sirah.* It is, of course, a secondary source, originally transmitted orally, and any critical analysis must remain aware of this. However, orthodox Muslims accept it as broadly accurate. It includes, supposedly, Muhammad's exact words, his teachings, his silent permissions or disapprovals, his habits, his life story, and reports about his closest companions. This, of course, means that in a context in which Muhammad is a paradigm of humanity, we have a great deal of detail, so orthodox Muslims believe, with regard to what Muhammad thought about things and how, precisely, you should live if you are a good Muslim.

To a certain extent, this makes Islam unique among the world's major religions; it conceives of a "Perfect Man" and it has a store of historical information about him, or believes that it has. With Jesus, there is a much greater extent to which the "Christ of Faith" is key, because so little is known about Jesus, the carpenter turned prophet.11 This renders Christianity malleable in a way that is far less true of Islam. If you ask, "What Would Muhammad Do?" then you have multiple texts where you can probably find the answer, or something reasonably approaching an answer. If Christian theologians ask, "What would Jesus have thought of homosexuals?", a pertinent issue in the modern Church, then there's considerable room for debate. Jesus doesn't mention homosexuality at all. You could argue that he was a man of his time and a committed

11—C. Stephen Evans, *The Historical Christ and the Jesus of Faith: The Incarnational Narrative as History.* (Oxford: Clarendon Press, 1996).

Jew. Judaism forcefully condemned homosexuality, with Jewish texts requiring that homosexuals be executed. Saint Paul was supposedly guided by the Holy Spirit, and he condemned homosexuality.12 Therefore, Jesus would have regarded homosexuality as abhorrent and sinful. On the other hand, Jesus specifically argued that he was overturning Jewish Law; reducing everything down to the two commandments of obeying God and loving your neighbor. He consorted with those who were marginalized, such as prostitutes and lepers, and he preached that the marginalized, of which, it might be argued, homosexuals are an example in many societies, would inherit the Earth. Certainly, many Christian theologians have attempted to argue that being a practicing homosexual is consistent with being a follower of Jesus.13 By contrast, when it comes to Muhammad's view on this issue, there's rather less room for maneuver:

> *The Prophet of Islam said: "Whoever you find committing the sin of the people of Lot, kill them, both the one who does it and the one to whom it is done."*
> —Al-Tirmidhi, Suna, 1376

> *The Prophet cursed effeminate men and those women who assume the manners of men and he said, "Turn them out of your houses!"*
> —Ibn Abbas, 7: 72, 774

12—R. R. Reno, "Redemption and Ethics," In *The Oxford Handbook of Theological Ethics*, eds. Gilbert Meilaender and William Werpehowski (Oxford: Oxford University Press, 2005), 29.
13—Matthew Vines, *God and the Gay Christian: The Biblical Case in Support of Same-Sex Relationships* (New York: Crown Publishing, 2014).

It should be emphasized that there is some debate among Islamic theologians, insomuch as Muhammad once protected an effeminate man from being lynched and employed a feminine man as a servant. But, clearly, this is not quite the same thing as condoning homosexuality. The *Koran,* these parts of the *Hadith*, and the immediate followers of Muhammad are so clear regarding the unacceptability of homosexuality that orthodox Muslims accept that homosexuality, on these grounds, was condemned by the Prophet.[14] Homosexuality is *haram* (not permissible) rather than *halal* (permissible). A liberal Islamic theologian has presented the fudge that the *Koran* says nothing *explicit* about homosexuality,[15] but it can be countered that other Islamic texts, based on an oral tradition of the Prophet Muhammad's teachings, are very clear about it. And, of course, there are many issues, such as the requirement to fast, over which there is simply no dispute at all. It is true that liberal Islamic theologians have emphasized, as have liberal Christian theologians with regard to Biblical texts, that Islamic texts need to be understood in their context and reinterpreted for different contexts. The most liberal Islamic theologians aver that the only clear point that can be discerned from the *Koran* is that Muhammad received some kind of revelation about moral values. The specifics are merely a matter of the context in which

14—See, however, Scott Alan Kugle, *Homosexuality in Islam: Critical Reflection on Gay, Lesbian, and Transgender Muslims* (Oxford: One World Books, 2010).
15—Kecia Ali, *Sexual Ethics and Islam: Feminist Reflections on Qur'an, Hadith and Jurisprudence* (London: Oneworld Publications, 2016).

Muhammad was operating.[16] The leading Islamic Modernist theologian Muhammad Abduh (1849-1905), an Egyptian, argued, in effect, that the greatest gifts that Allah bestowed upon humanity were independent will and independent thought, and with these gifts happiness could be attained. Islamic texts were allegorical or merely of their time.[17]

These views are very far from orthodox Islam, most obviously because the *Koran* is clear that it is to be taken at face value and that the Doctrine of Abrogation is to be applied whenever there is any confusion. Interpreting the Bible is, therefore, much more open to debate, meaning that parts of it can more legitimately be interpreted as metaphorical if this is how the reader wishes to interpret them. But, even so, "Christian Non-Realism" is very far from orthodox Christianity. Espoused by Anglican priest the Rev. Don Cupitt, its doctrine seems to be (it tends to be expressed through ambiguous metaphors) that religion should be defined as an optimistic attitude to the cosmos. God doesn't exist but you should live as if He does and thus try to be moral, like Jesus told you to be, and pray to the "dead God." Doing this makes you a "Christian."[18] There are so many Christian theologians who by any reasonable definition are atheists. They redefine "God" as "depth" or "love" and they will often try to argue that this definition is so much more sophisticated and profound than that of

16—Intisar A. Rabb, "Ijtihād" In *The Oxford Encyclopedia of the Islamic World*, ed. John Esposito (Oxford: Oxford University Press, 2009).
17—James L. Gelvin, *The Modern Middle East* (Oxford: Oxford University Press, 2008).
18—Ronald T. Michener, *Engaging Deconstructive Theology* (London: Routledge, 2016), 144.

religious conservatives.[19] It is as though those who hold such views are simply atheists who have an emotional attachment to their childhood religion and thus want to innovate a clever way in which they can square the circle of their conflicted sense of identity. And perhaps they also want to persuade people, via superficially being able to justify something so seemingly contradictory, that they are highly intelligent and profound, something that has been termed "clever silly."[20] But such people are, clearly, not only fringe elements, but logically incoherent: redefining commonly understood words out of all recognition such that they can justify as logically coherent a fractured, conflicted sense of self that is both "religious" and "atheist." For orthodox Muslims, God exists; Muhammad is His Prophet; the *Koran* is divine revelation; the *Koran* is clear how any internal contradictions should be dealt with; and the *Sunna* is broadly reliable, though there are disputes in certain areas.

As we will now see, many key Islamic exhortations actually have the effect of reducing intelligence. Though, as we will also discover, there is a positive dimension to this in terms of winning in the battle of group selection. They also indirectly increase ethnocentrism. We will begin with the essence of Islam: Its Five Pillars.

19— For example, Robinson, *Honest to God*, *op. cit.*

20—See Edward Dutton and Dimitri Van der Linden, "Who Are the 'Clever Sillies'? The Intelligence, Personality, and Motives of Clever Silly originators and Those Who Follow Them, *Intelligence* 49 (2015): 57-65; Bruce G. Charlton, "Clever Sillies: Why High IQ People Tend to be Deficient in Common Sense, *Medical Hypotheses* 73 (2009): 867-870.

Chapter 4

Do the Five Pillars of Islam Reduce Intelligence?

Shahada—Faith

The five pillars of Islam, around which the religion is based, are Faith, Prayer, Charity, Fasting, and making a pilgrimage to Mecca, the birthplace of Muhammad, at least once in your life time. None of these are particularly good for your intelligence and the majority are bad for it. We will look at them in their order of importance, beginning with faith.

You must "believe" that Allah is the one true God and that Muhammad is his prophet, as well as further more specific beliefs about the nature of the *Koran* and other holy books. In other words, you must engage in what the American psychologist Kevin MacDonald has termed "effortful control."[21] You must persuade yourself not to think in a critical manner with regard to certain key aspects of life, in order that you can remain a member of the Muslim community and psychologically cope with so being. People use this tactic all of the time. It is abundantly clear from the evidence that there exist genetic racial differences in intelligence and that evolutionary variation is the simplest explanation. However, many psychologists, knowing that their lives will become difficult if they admit this but also believing in the importance of logic and empirical evidence, use effortful control to convince themselves that the evidence does not prove what it so manifestly does prove. Muslims are simply asked to do the same thing. It would follow, in such an environment, that people would practice analytical thinking to a lesser extent, causing them to be less adept at analytical thinking. This same process would happen to the society more broadly, meaning that it would not value analytical thinking in the way that less religious societies would. This would potentially permeate every aspect of society, meaning that the education system would be less focused on analytical thinking and more focused on learning information by rote, as tends to be the case with education systems in Islamic countries[22] and was the case in Western countries when they were more religious.[23] The education system would also be more focused on religious instruction.

21—Kevin MacDonald, "Effortful Control, Explicit Processing, and the Regulation of Human Evolved Predispositions," *Psychological Review,* 115 (2008): 1012-1031.

22—Dutton, "A Negative Flynn Effect in Kuwait," *op cit.*

23—Flynn, *Are We Getting Smarter? op. cit.*

National IQs have been collected by Richard Lynn and Tatu Vanhanen.[24] They have been found to strongly correlate, at around 0.8, with national differences in other measures of cognitive ability, such as student assessment tests; so they are clearly highly robust. Lynn's entire national IQ dataset has been reanalysed from scratch, so eliminating all of the sub-standard samples and other anomalies for which his data-set has been criticized.[25] The results have then been recalculated by German political scientist David Becker. Becker's results correlate with Lynn and Vanhanen's at 0.87.[26] Thus, Lynn and Vanhanen's national IQ scores are highly reliable, Becker maintains an online database called *View on IQ* where the skeptic can check all the sources. A summary of Becker's results has also been presented as *The Intelligence of Nations.*[27] There is no question about it. There are meaningful differences in average IQ between nations.

There is a symbiotic relationship between economic growth and national average IQ. Low IQ countries display poor economic growth because intelligence predicts the ability of a nation to become wealthy. But poor economic growth also limits intelligence, by failing to bring that

24—Richard Lynn and Tatu Vanhanen, *Intelligence: A Unifying Concept for the Social Sciences* (London: Ulster Institute for Social Research, 2012).

25—Earl Hunt and Robert J. Sternberg, "Sorry, Wrong Numbers: An Analysis of a Study of a Correlation Between Skin Color and IQ, *Intelligence,* 34 (2006): 131-137.

26—David Becker, D, "A Warm Welcome, *View on IQ,* http://viewoniq.org/?p=41 (accessed August 15, 2019); Richard Lynn and David Becker, *The Intelligence of Nations* (London: Ulster Institute for Social Research, 2019).

27—Lynn and Becker, *The Intelligence of Nations, op. cit.*

nation's IQ to its phenotypic limit: schooling will be of poor quality, fewer people will be literate and those of high intelligence will be less able to find other people like them, due to deficient transport and communications. People's intellectual development may also be stunted by insufficient nutrition and periods of childhood illness, and the nation is less likely to achieve a minority with particularly pronounced intelligence, something that has been shown to be especially important in terms of economic growth. If anything, the presence of a high IQ class, a "smart fraction," is more significant than the average IQ of the nation.[28] In his book *Cognitive Capitalism,* German psychologist Heiner Rindermann has demonstrated that in poorer countries, such as Islamic countries, level of wealth positively affects its cognitive ability, seemingly by raising intelligence to its phenotypic limit via a functioning educational system and access to proper nutrition.[29] However, the richer a country is, the more likely this is to have been reached, so the more its cognitive ability positively impacts level of wealth. The smarter it is, the smarter will be its cognitive elite. And the smarter they are, the better the nation will be organized, the more inventive it will tend to be and the wealthier it will become.[30]

So the reduction in analytical thinking brought about by the focus on faith in Muslim countries would

---

28—Heiner Rindermann, *Cognitive Capitalism: Human Capital and the Well-Being of Nations* (Cambridge: Cambridge University Press, 2018).
29—*Ibid.*, 232.
30—See Edward Dutton, Review of *Cognitive Capitalism*, by Heiner Rindermann. *Journal of Social, Political and Economic Studies,* 43 (2018): 351-354.

create an amplification effect. With the education system not pushing analytical thinking to its phenotypic maximum, the society would produce fewer per capita important inventions; it would be less able to operate such innovations when imported from more analytically-oriented societies; it would be materially poorer. This would mean that a higher percentage of people would be in poverty; failing to reach anywhere close to their phenotypic IQ maximum due to disease, poor nutrition, and food shortage. This would be amplified further by civil strife and other internal conflicts, making people act in an instinctive, and so not analytical, fashion, further reducing the extent to which they employed their analytical faculties. Indeed, Rindermann[31] has highlighted evidence that war reduces intelligence. In a British cohort born in 1921 and tracked throughout their lives, intelligence was a key predictor of having been killed in World War II. High IQ people, with their community-oriented, altruistic values, really were more likely to have got killed in World War II, and presumably World War I. More likely to become officers, they would be at the vanguard of any attack, but even the most intelligent among the troops would have been more prone to gallant, self-sacrificing behavior.[32]

So, this focus on the importance of accepting certain doctrines, whatever reason or empirical evidence might tell you, is wont to indirectly negatively impact the society's average intelligence. However, there is an obvious plus side. As we have already noted, religiousness makes people and groups more ethnocentric. Poverty

31—Rindermann, *Cognitive* Capitalism, *op. cit.*, 243.
32—See Dutton, Review of *Cognitive Capitalism, op. cit.*

and political instability induce stress and stress makes people more religious and so more ethnocentric. In a globalized world, people in poor countries are aware of their relative poverty and they are also confronted with an elite that enjoys Western standards of living. This is likely to create feelings of exclusion, which will in turn elevate religiousness, which will in turn increase ethnocentrism. Poverty also makes people more instinctive, something which will heighten both religiousness and ethnocentrism. Finally, people don't tend to want to migrate to poor countries. It has been shown that when foreigners come and live in a community, it reduces trust even among the natives, because there is now a group to whom fellow natives can defect.[33] Indeed, Tatu Vanhanen showed in his book *Ethnic Conflicts*[34] that the more multicultural a society is, the lower are its levels of generalized trust. This is not generally a problem in mono-racial religious societies, argues Vanhanen, because they tend to be poor meaning that foreigners are disinclined to migrate to such societies.

By contrast, the high standards of living that exist in Western countries mean the level of stress experienced by Western people is relatively low, something that will reduce the extent of their religiousness and so reduce the extent of their ethnocentrism. Western societies are unlikely to feel a strong sense of exclusion, and they will act as a magnet for foreigners, thus reducing trust, and thus positive ethnocentrism, among the natives. Low in stress,

33—Robert Putnam, "*E Pluribus Unum*: Diversity and Community in the Twenty-first Century," The 2006 Johan Skytte Prize lecture, *Scandinavian Political Studies* 30 (2007), 137–174.

34—Tatu Vanhanen, *Ethnic Conflicts: Their Biological Roots in Ethnic Nepotism* (London: Ulster Institute for Social Research, 2012).

Western people will also be low in instinct, meaning it will take a great deal to trigger their ethnocentric impulses. Accordingly, we can begin to see how Islam, by repressing the intelligence of its adherents, elevates its adherents' ethnocentrism. This means that when the ethnocentrism of a rival group is particularly low, Islam can dominate it even if that rival has higher average intelligence and superior technology.

The Islamic philosopher Ibn Khaldun (1332-1406), who was from what is now Tunisia, conceived of what he called *asabiyah,* meaning solidarity or group feeling, essentially ethnocentrism. He argued that living in harsh desert conditions elevated *asabiyah,* allowing cities and thus civilizations to be built. However, due to the luxurious conditions available in these cities, at least to the elite, people inevitably became decadent and *asabiyah* levels declined. Eventually, a new group of desert tribesmen, high in *asabiyah,* would arrive and they would successfully conquer the civilization, and the process would begin all over again, as these new desert tribesmen built a civilization and their *asabiyah* declined accordingly.[35] High *asabiyah* seems to reflect poverty, stress, and religiousness. By reducing intelligence to an optimal low, Islam elevates *asabiyah* to an optimal high, sometimes sufficient to overthrow a highly intelligent and developed civilization.

35—Allen James Fromherz, *Ibn Khaldun, Life and Times.* (Edinburgh: Edinburgh University Press, 2010).

Salah—Prayer

Good Muslims must pray to Allah five times each day, facing Mecca. The negative consequences of this for the economy, especially in a globalized context, are obvious. The first call to prayer is very early in the morning, which means that people's sleep is interfered with. One of the most frustrating things when I was in Alanya in southern Turkey in 2008 was being awoken in the morning by the call to prayer. Hauntingly beautiful though it sounds, it was broadcast from all mosques around the town through loud speakers. In the week I was on holiday there, I never managed to sleep through it. Compared to those who are not awoken in the early hours of the morning, Muslims will have less deep sleep, which means they will have less time processing and practicing important aspects of life, this being the evolutionary purpose of dreams, in line with experimental evidence that even animals dream.[36] It follows that they will perform any task allotted to them sub-optimally when compared to those who have had sufficient time to dream. An important aspect of sleep, in general, is allowing the body to rest and repair itself. This process will be interfered with in observant Muslims, leading to their being tired and their brains not functioning optimally. This is a central aspect of intelligence due to its association with quick reaction-times, which reflect the fact that, to a certain extent, intelligence can be reduced down to a high functioning nervous system.

36—See Katja Vallia, Antti Revonsuob, Outi Pälkäsc, Kamaran Hassan, Ismaild Karzan, Jalal Alid, Raija-Leena Punamäki, "The Threat Simulation Theory of the Evolutionary Function of Dreaming: Evidence from Dreams of Traumatized Children, *Consciousness and Cognition* 14 (2005): 188-218.

In addition, prayer lasts between about five and fifteen minutes. These are periods of time, during the day at least, where work could be getting done and where analytic thought could be being engaged in. But it is not. As such, it will have a negative effect on the economies of Muslim countries, indirectly causing a worse education system and greater poverty. And this will be compounded by people being, relative to other societies, tired and insufficiently practiced in important tasks. So the system of praying five times a day will have the indirect effect of reducing the nation's IQ and reducing the nation's productivity and, by extension, its IQ. By leaving them tired, with insufficient deep sleep, less able to rehearse important tasks and with less time to engage in analytical thought, the system will also reduce the performance IQ of individual Muslims. This, unlike "Faith," is a problem unique to Islam.

But, again, there are plus sides to this particular intelligence-reducing technique. There is a degree to which how often you pray can be monitored; such as by your work colleagues or your family members. It thus allows other Muslims to assess how genuinely committed to Islam you are. If they are confident that you are genuinely committed, then they are better able to trust you, trust being a central aspect of positive ethnocentrism. It also permits them to better understand who should not be trusted, by observing who is not responding to the call to prayer. So, *salah* helps to elevate trust and thus positive ethnocentrism. It demands a sacrifice of time and energy, driving out those who are not prepared to make this sacrifice, exposing them as the free-loaders and non-co-operators that they are. It also means that the entire society is undergoing a shared ritual at the same time, elevating the degree to which people feel similar to each other. People tend to bond with and trust those who are similar to themselves. Indeed, when we regard someone as trustworthy, we unconsciously rate their face as being similar to ours, even if it objectively is not.[37]

Prayer also elevates religiousness. Religious experiences occur when a person is, in effect, in a highly instinctive state. This happens when they are highly stressed, which is why dramatic conversion experiences tend to happen at times of stress, such as among late adolescent university students, away from home for the first time. As an undergraduate, I was at Durham University in the northeast of England. It is a kind of mini or wannabe Oxford. Like Oxford, a very large proportion of its students have been to public schools (highly prestigious private secondary schools where pupils usually board). Six people in my fifteen-person hall of residence in my first year had been to public schools, two of those to Eton. Durham University maintains all kinds of public school traditions, such as formal meals. Many of the students were rejects from Oxford or Cambridge.[38] While I was there, I knew a number of people who underwent religious experiences. They were homesick, away from home for the first time, away from the affluent south of England for the first time, and disoriented by meeting people from very different backgrounds. This left them questioning who they

37—Harry Farmer, Ryan McKay, and Manos Tsakiris, "Trust in Me: Trustworthy Others Are Seen as More Physically Similar to the Self," *Psychological Science* 25 (2014): 290-292.
38—Edward Dutton, *Meeting Jesus at University: Rites of Passage and Student Evangelicals* (Aldershot: Ashgate, 2008).

were, which left them stressed, and it is this that can lead to religious experiences. As a consequence, they became evangelical Christians, at least for periods of time. In one case, the conversion happened in the shower of my own hall of residence, with the girl who experienced it declaring, "I know my place in Heaven is secure."[39] However, the other means of inducing a religious experience is through intense calm, such as meditation or prayer. In this state, people are highly prone to their cognitive biases and, thus, feeling that God is present.[40] In addition, we have already observed that poverty elevates religiousness and ethnocentrism and that less intelligent people are more religious. The economic and psychological damage that *salah* will do to Muslims may be more than compensated for by the extent to which it indirectly galvanizes them in terms of ethnocentrism.

Zakat—Charity

Observant Muslims, especially the majority Sunni, must give one fortieth of their savings to the needy. This is believed to create social solidarity among the Muslim community and reduce tension between the rich and the poor. Socioeconomic status is around 70 percent genetic across generations,[41] it is robustly correlated with intelligence, and intelligence is strongly genetic.

39—*Ibid.*
40—See Andrew Newberg, Eugene G. d'Aquili, and Vince Rause, *Why God Won't Go Away: Brain Science and the Biology of Belief* (New York: Ballantine, 2002).
41—Gregory Clark, *The Son Also Rises: Surnames and the History of Social Mobility* (Princeton: Princeton University Press, 2014).

Thus, such a practice is simply helping those with low intelligence to pass on their genes and so reducing the intelligence of Muslim societies. On the other hand, this has to be balanced with the likely fact that it will reduce social tension and criminality, thus potentially reducing stress, and allowing the economy to work more efficiently. So, there does not appear to be a particularly persuasive case for arguing that *Zakat* reduces intelligence, though it is possible that it may do overall.

It is, however, likely to boost levels of ethnocentrism, by virtue of elevating a sense of solidarity, a sense of everyone being the same. We have already observed Farmer and his colleagues' finding that we associate similarity to self with trust.[42] If socioeconomic differences within a society are substantial, there is a tendency for people not to perceive those whose lifestyles are very different as part of their in-group. The very poor are often regarded with feelings of disgust; people react to them similarly to the way in which they react to images of people with skin diseases[43] while the very wealthy are often regarded with contempt and envy.[44] *Zakat* at least has the potential to reduce this problem.

42—Farmer, H., McKay, R. & Tsakiris, M. (2014). Trust in Me: Trustworthy Others Are Seen as More Physically Similar to the Self. *Psychological Science*, 25: 290-292.
43—Bernice Lott, "Cognitive and Behavioural Distancing from the Poor," *American Psychologist* 57 (2002): 100-110.
44—Suzanne Horwitz and John F. Dovidio, "The Rich—Love Them or Hate Them? Divergent Implicit and Explicit Attitudes Toward the Wealthy," *Group Processes and Intergroup Relations*, 20 (2015): 3-31.

Sawm—Fasting

Muslims are expected to fast during the Holy Month of Ramadan, a celebration of the month during which they believe that the *Koran* was revealed to the Prophet Muhammad by Allah. During this month, Muslims are not supposed to eat between sunrise and sunset. As with many religious practices, there are clear evolutionary benefits to this. There are benefits to being a Muslim: you are part of a relatively ethnocentric group that will, to a certain extent, look after you. In return for this, the group needs to ensure that all of its members are genuinely committed, that they are not freeloaders. It needs evidence, in other words, that members are sufficiently rule-following and cooperative that the group will remain high in ethnocentrism and thus highly competitive. As we have already discussed, they can achieve this through tests of membership, by forcing members to make certain sacrifices as proof of their commitment to the group. You can fake praying and pretend that you believe in Allah, but it is much more difficult to get away with eating during the day during Ramadan without being noticed. This is the point of asking them not to eat all day during Ramadan. It helps to expose and drive out the non-co-operators, who can then be shunned and effectively purged from the group's gene pool. Those who follow the rules, thus signaling their altruism and cooperativeness, will likely be deemed more attractive potential mates and friends: they can be trusted; they are true followers of the Prophet. Those who fail to follow the rules will be less attractive; they may even become outcasts, with other Muslims not wanting to marry them. In these key ways, *sawm* helps to make the group more competitive.

When I was aged six and seven, my best friends at Infants School were Sa'ad, whose parents were from Iraq, and Yusuf, whose parents were South Asians from East Africa. Sa'ad's parents dressed like Westerners and were seemingly very liberal. His religion only came to light one Sunday in Spring 1987 when he came round to play. My parents took us to the local pub so that Sa'ad and I could play in the beer garden, where there were swings and a slide, and my parents could become a bit merry. In those days, English pubs were only just beginning to serve food. The Royal Oak's fayre was beef or ham rolls. My father bought some and I took the last beef roll, leaving Sa'ad with the ham one. Sa'ad, a highly extroverted boy, seemed suddenly nervous. He didn't know why, but he just wasn't allowed to eat that; so my father had to go back to the bar and swap it for another beef one. Yusuf's parents, qualified accountants who ran a hardware shop, were far more conservative. The father had a long beard and wore a *kufi* (Islamic white hat) and the mother, who was a superb cook, wore a hijab, taking it off inside. When I ran into Yusuf in the year 2000, he had adopted the same style as his father. I vividly remember, from about 1988, that Yusuf had to fast during Ramadan. I can see the miserable, slightly irked look on his face as myself and others did all that we could to persuade him to eat. Doubtless his hunger would also have reduced his performance IQ, as it has been shown that hungry children perform less well in mathematics tests than similar non-hungry children.[45] It would follow that Islamic societies as a whole would think

45—Chad Cotti, John Gordanier, and Orgul Ozturk, "When Does Count? The Timing of Food Stamp Receipt and Educational Performance, *Economics of Education Review,* 66 (2018): 40-50.

less logically during Ramadan, damaging the country's economy, elevating poverty levels, and further damaging cognitive development via less stimulating environments and worse nutrition.

However, this pillar also has the effect of reducing the intelligence of Muslims in a more long-term way. A fascinating literature review of the economics of Muslim countries has found that the Ramadan fast has a number of effects that indirectly reduce intelligence.[46] Fasting depresses a country's production and, therefore, its economic growth. Accordingly, the Muslim practice of fasting indirectly reduces IQ in Muslim countries, as we have noted the importance of economic status to national IQ and *vice versa*. The same literature review reveals a number of studies which have found that fasting in early pregnancy, as a woman in early pregnancy may not know she is pregnant, results in children having "shorter lives, worse health, less mental acuity, lower educational achievement, and weaker performance in the labor market."[47] One study reviewed found that 15 percent of mental disability among Muslims can be explained by their mothers fasting in very early pregnancy. Though the Koran states that if you are ill during Ramadan you should postpone your fast until later, it says nothing about pregnant women doing so. This practice is likely to have a knock on effect. In extreme cases it causes mental disability in the child but in many others it will simply mean that the child's IQ is not as high as it otherwise

46—Timur Kuran, "Islam and Economic Performance: Historical and Contemporary Links," *Journal of Economic Literature,* 56 (2018): 1292-1359.
47—*Ibid.*, 8.

would have been. And it must be kept in mind that, unlike with Christianity, these practices are very specifically set out in sanctified literature.

Hajj—Pilgrimage to Mecca

The least important of the five Pillars of Islam is that you must make a pilgrimage to Mecca at least once in your life time. In a context in which the standard of living in Muslim countries is considerably lower than in Western countries, this may well indirectly negatively impact both individual and national IQ, the latter, in turn, negatively impacting individual IQ. Unless you live close to the city, it is expensive to go on a pilgrimage to Mecca. This is money that might be better spent on more nutritious food, better quality education for one's children, and books and other intellectually stimulating material. *Hajj* will also involve taking a period of unpaid leave, at least in many Muslim countries, which will also have a negative impact on finances. Furthermore, if you go on the pilgrimage you will be with other highly observant Muslims, especially considering that *Hajj* is the least important of the five Pillars of Islam. Controlling for other factors, such as wealth, we would expect these highly religious people to have lower IQ than those who do not go on the pilgrimage. And, as we have seen, spending time with low IQ people reduces your own IQ. So, for some pilgrims at least, the pilgrimage would potentially have a negative impact on intelligence, albeit only in a small way.

There is evidence that Pakistanis who go on *Hajj* display attitudes that are more beneficial to the economy than those who do not. They are more likely to accept female education and employment, more tolerant towards foreigners and non-Muslims, and more tolerant to members of different ethnic groups within Pakistan. However, this cannot be as a result of interacting with foreigners on the pilgrimage, because evidence indicates that people strongly stick with their own group while they are on the pilgrimage.[48] Thus, the simplest explanation is that those who can afford to go on the pilgrimage are more intelligent and, therefore, are relatively low in conservative attitudes by the standards of observant Pakistani Muslims. The practice of *Hajj* does not assist the Pakistani economy. It is likely to make it less efficient, with all that this will indirectly do to average IQ in Pakistan and, by extension, other Muslim countries.

What *Hajj* does permit you to do, however, is engage in a kind of "virtue-signaling." It is a means through which you can stress how religiously committed you are, allowing you to emphasize such attractive qualities as devotion, cooperativeness, and rule-following. These traits are actually positively associated with socioeconomic status through the personality trait Conscientiousness, the essence of which is rule-following and impulse control.[49] The fact that you, or even a close relative, have been on *Hajj* will thus make you attractive to potential partners, attaining a partner being extremely competitive when

48—*Ibid.*

49—Daniel Nettle, *Personality: Understanding What Makes You Who You Are* (Oxford: Oxford University Press, 2007).

you are a male in a society that practices polygamy, as orthodox Islamic societies do. Accordingly, there may be a small degree to which the bragging rights associated with *Hajj* elevates the extent to which the most devout are able to procreate, though far more research would need to be conducted to prove this.

The Five Pillars of Islam and Non-Muslim Intelligence

We can conclude that at least three of the five pillars of Islam directly or indirectly reduce intelligence. Fasting causes economic inefficiency due to tiredness and hunger, with all that stems from that in terms of the educational system, poverty, disease and so on. It can also negatively impact the cognitive development of children. "Faith" reduces the practice of analytical thinking, damaging IQ directly and indirectly by negatively impacting the entire economy. "Prayer," likewise, leads to economic inefficiency and suboptimal IQ at the individual level due to tiredness and lack of deep sleep. The impacts of pilgrimage and especially alms-giving are, at best, neutral, and there is a not unconvincing case for arguing that the impact of pilgrimage is negative. Put simply, following the five Pillars of Islam, which are fundamental to the faith, reduces intelligence.

And it doesn't just reduce intelligence for Muslims. If there is large scale immigration of Muslims, with their significantly lower IQ, into Western countries, then this will manifestly reduce the average IQ of the country in question, as has been shown elsewhere.[50] But it won't do this solely by virtue of the presence of a low IQ block in the otherwise Western population. The effect will be far more insidious than that. Cultural differences and the simple presence in schools of demanding and low IQ students will make teaching *all* children less efficient. It will reduce the ability of the education system to create an optimally intellectually stimulating environment for *all* children, certainly for those who have to share schools with these low IQ immigrants. More and more natives will have to interact with them, where they would have been interacting previously with those of higher intelligence. This will help to reduce the native IQ as well. As the national IQ falls, the entire country will start to become less efficient, more stressful to live in and less able to maintain the population's IQ at its phenotypic maximum. Therefore, the IQ of everyone, even the natives, will begin to fall. Ethnic conflict is likely to develop in these circumstances. The ability to think in a rational way is, as we have discussed, reduced by stress and conflict and war tends to cause dire stress. Detailed research by the Finnish political scientist Tatu Vanhanen (1929-2015) has shown that the more ethnically diverse a country is, the more ethnic conflict there will inevitably be.[51] Drawing on data from across the world, Vanhanen finds that the degree of conflict and warfare within a country correlates with its level of ethnic diversity at 0.66. Islam reduces intelligence not just for adherents but for everybody who lives in a country where there is a Muslim minority of any numerical impact.

50—Lynn, *Dysgenics, op. cit.*
51—Vanhanen, *Ethnic Conflicts, op. cit.*

However, crucially, just as three of the Pillars of Islam reduce intelligence, they also increase ethnocentrism, for Muslims. For the Westerners among whom they come, it is rather more complicated. On the one hand, they will help to reduce the intelligence of society and increase levels of stress, which may elevate negative ethnocentrism, leading to ethnic conflict. But, on the other, their presence is likely to reduce trust levels among the native population, and so decrease levels of positive ethnocentrism as well. Accordingly, Islam's impact on Western ethnocentrism can be regarded as ambiguous. Islam's effects on its followers, however, should not be in doubt.

Chapter 5

Purdah and Marriage Patterns

Female Genital Mutilation

Having explored the Five Pillars of Islam, we will now turn to religiously mandated Islamic traditions. Sharia Law is a system of Islamic Law that is based around the recorded teachings of the Prophet Muhammad. It is a strongly patriarchal system in which females are oppressed from a very young age. As we will see below, there are many evolutionary advantages to this system. However, it is not especially advantageous in terms of elevating Muslim average intelligence. Indeed, it is disadvantageous.

In the *Hadith*, the Prophet Muhammad is recorded as having exhorted his followers, a number of times, to circumcise their children:

> *Abu al- Malih ibn 'Usama's father relates that the Prophet said: "Circumcision is a law for men and a preservation of honor for women.*
> —Ahmad Ibn Hanbal 5:75; Abu Dawud, Adab 167

> *Narrated Umm Atiyyah al-Ansariyyah: A woman used to perform circumcision in Medina. The Prophet (peace be upon him) said to her: "Do not cut severely as that is better for a woman and more desirable for a husband."*
> —Sunan Abu Dawud 41:5251

Young Muslim girls, according to Muhammad, are to be circumcised. There are clear advantages to this in terms of ethnocentrism. If the girl has been circumcised, she will be less able to enjoy sex. This will mean that she is less likely to commit adultery and so less likely to cuckold her husband. This, in turn, means that there is a greater incentive for her husband to invest both in her and in his children by her, because his paternity anxiety is reduced. The family that choose to circumcise their daughter are, in doing so, displaying their religious commitment and so emphasizing that they are cooperative members of the community, an act which will help increase their status in the community. If it is known that the daughter has not been circumcised, then she is likely to be unable to find a husband or her husband will shun or kill her on her wedding night when he realizes that she has not been circumcised. So circumcising her certainly helps her to pass on her genes in this context.[1] More generally, it increases cooperation in the society as a whole, because it means that men will be more willing to invest in their families and less likely to get into jealous fights with other men, because their level of paternity anxiety will be low. Accordingly, they will be able to trust other men, and

1—Yael Sela, Todd K. Shackelford and James R. Liddle, "When Religion Makes It Worse, *op. cit.*

their wives, more, helping to foster a more cooperative and stable society.

In Somalia, this is taken to an extreme, perhaps because the country has a relatively low IQ even by the standards of Muslim countries; it is around 70. This means that innate trust levels are extremely low, so radical measures are required to induce men to trust that their wives are virgins when they marry them and they are most unlikely to commit adultery, since they are unlikely to enjoy sex. As we will see shortly, a Somali girl is circumcised, after which her vagina is sewn up, to be torn open by her husband on their wedding night. A tiny hole is left for the girl to urinate through, meaning that she must spend a very long time in the loo in order to empty her bladder.

Female circumcision is particularly prevalent in Muslim areas of Africa. According to UNICEF data from 2013, 96 percent of Somali females aged between 15 and 49 have been circumcised, the most of any country. In Sudan it is 88 percent, in Egypt it is 91 percent, in Yemen it is 23 percent, though in Iraq it is just 8 percent.[2] Rates of Female Genital Mutilation (FGM) tend to be higher in poorer regions of countries, which also tend to be more religiously conservative. Thus, in southern Egypt, in the years 2014 to 2015, it was found that 92 percent of women between the ages of 15 and 49 had been circumcised. However, in the wealthier northern part of the country it was 86 percent.[3] The negative side of this tradition is that being circumcised, something

2—The World Health Organization, "Female Genital Mutilation (FGM)," http://www.who.int/reproductivehealth/topics/fgm/prevalence/en/ (accessed August 15, 2019).

3— Dutton, "Regional Differences in Intelligence in Egypt," *op. cit.*

that is done at a relatively young age, though not as a baby, is extremely traumatic for Muslim girls. In some ways, it is the end of their innocence. In her memoir, *Infidel*[4] the Somali refugee and Dutch politician Ayaan Hirsi Ali (b.1971) recalls being circumcised at the age of five: "Little girls are made 'pure' by having their genitals cut out . . . the practice is always justified in the name of Islam. Uncircumcised girls will be possessed by devils, fall into vice and perdition, and become whores. Imams never discourage the practice: it keeps girls pure." Ali remembers that:

> *Then the scissors went down between my legs and the man cut off my inner labia and clitoris. I heard it, like a butcher snipping the fat off a piece of meat. A piercing pain shot up between my legs, indescribable, and I howled. Then came the sewing, the long, blunt needle clumsily pushed into my bleeding outer labia, my loud and anguished protests, Grandma's words of comfort and encouragement . . . When the sewing was finished the man cut the thread off with his teeth. That is all I can recall of it.*[5]

She then had to lie perfectly still for a week with her legs tied together, so that the sewn up flesh would fuse, her legs untied only for her to urinate, a process that was excruciating. After the week in bed was over, the itinerant circumciser, part of "the blacksmith clan," returned to remove the stitches, a process that was obviously extremely painful. Almost 20 percent of Somali girls who

4—Ayaan Hirsi Ali, *Infidel: My Life* (London: Pocket Books, 2007), 37.
5—*Ibid.*, 32.

are circumcised report symptoms consistent with Post-Traumatic Stress, evidently related to vivid memories of their genital mutilation.[6] In addition, it is common for these girls to experience less severe forms of lasting trauma.[7] Ayaan Hirsi Ali's sister was obviously traumatised by her circumcision. Ali recalls the change in her sister's behavior: "My once cheerful, playful little sister changed. Sometimes she just stared vacantly at nothing for hours."[8] Childhood trauma is not helpful in terms of IQ. Among those who are psychologically normal, as opposed to those with psychotic disorders, childhood trauma can decrease adult IQ by up to 6 points, by virtue of it causing the brain to develop in a sub-optimal manner.[9] Put simply, FGM contributes, if only in a small way, to making females markedly less intelligent; and it will lead to an amplification effect of sub-optimal intelligence spreading through Islamic societies. These sub-optimally intelligent women will raise their own children, and they will spend their early years, when the brain is particularly malleable, with women who will sub-optimally intellectually stimulate them, helping to make their own intelligence sub-optimal. Their husbands will spend a great deal of

6—Jeroen Knipscheer, Erick Vloeberghs, Anke van der Kwaak, and Maria van den Muijsenbergh, "Mental Health Problems Associated With Female Genital Mutilation," *BJPsych Bulletin* 39 (2015): 273–277.

7—Daniel Njoroge Karanja, *Female Genital Mutilation in Africa: Gender, Religion and Pastoral Care.* (Maitland, FL: Xulon Press, 2003).

8—Ali, *Infidel, op cit.*, 33-34.

9—Jim van Os, Anne Marsman, Daniela van Dam, Claudia J. Philippe Simons, GROUP Investigators, "Evidence That the Impact of Childhood Trauma on IQ Is Substantial in Controls, Moderate in Siblings, and Absent in Patients With Psychotic Disorder," *Schizophrenia Bulletin* 43 (2017): 316-324.

time with their wives, making their husbands' intelligence level lower than it would be had the wife not had her IQ interfered with via intense childhood trauma. So, female genital mutilation pushes IQ down both for those who experience it and, to a minor degree, those who associate with them.

Purdah and Intelligence

Attitudes to Muslim women also help to reduce IQ. Education is not considered to be particularly important for girls in many Muslim countries, so they are more likely than boys to be not sent to school at all, or to be withdrawn from school early.[10] As already discussed more broadly, this will have the effect of reducing their IQ and, by extension, reducing the IQ of their children and anybody else who spends a considerable period of time with them, especially at a formative age.[11]

From an evolutionary perspective, the veiling of fertile females makes a great deal of sense. In a relatively lawless society, and especially in one that is polygamous, women are in danger of rape, not least because there will be a large number of single males, the reasons for which we will discuss below. If females are subject to *purdah*, then they cannot sexually arouse these single males, because they will be covered up whenever they are in the presence of men and may also be chaperoned or in groups. This means that their husbands can be fairly certain that they

10—See Dutton et al., "Regional Differences in Intelligence in Egypt," *op. cit.*
11—Flynn, *Does Your Family Make You Smarter?*, *op. cit.*

have not been cuckolded and their fathers can be certain that the family's honor, and thus its socioeconomic prospects, is unlikely to be damaged by the daughter engaging in an extra-marital or non-marital affair. This, in turn, means that men are less likely to fight, more likely to invest in their children, and so, ultimately, better able to create a society where people invest in each other and trust each other.

To put it another way, *purdah* and also FGM, help to foster a slower Life History Strategy.[12] Men can trust that their children are their own, so they will trade mating effort, having sex with as many women as possible in the hope that some end up having your children, but investing very little in the individual children, towards nurturing effort. Having considerable confidence that the children are their own, they focus their energy towards nurturing these children via investment in them. This is possible because they are able to trust that these children are their own. Societies of this kind, precisely because they are high in trust, become cooperative and high in positive ethnocentrism. When combined with the high negative ethnocentrism that religiousness will tend to induce in them, you have a society that is highly adaptive in terms of group selection. In addition, if the level of education is controlled for, then *purdah* may actually have a positive impact on female intelligence. My own research group have found that among Saudi teenagers, male and female IQ are about the same in the Makkah region, this is the region that includes such cities as

12—J. Philippe Rushton, *Race, Evolution and Behavior*, *op. cit.*; Dutton, *J. Philippe Rushton: A Life History Perspective.* (Oulu: Thomas Edward Press, 2018).

Mecca and Jeddah. This is a religiously liberal area of Saudi Arabia, where *purdah* is not especially strongly enforced and where young females are relatively free. They are unlikely to cover their faces and may get away with not wearing a veil at all, the religious police simply overlooking the violation. By contrast, in Riyadh, which is strongly conservative, *purdah* is rigorously enforced. In Riyadh, teenage girls have higher IQ scores than teenage boys. A probable explanation is that the girls' lives are far more restricted than in Makkah, so they spend more time reading and doing school work, thus pushing their IQ to its phenotypic maximum.[13]

There is a probable downside to *purdah*. Muslim women who follow *purdah* do not go out very much and when they do, they are covered up. The result, and this has also been found among Orthodox Jewish samples, is Vitamin D deficiency.[14] It is certain that Vitamin D plays an important role in the development of a foetus' brain, meaning that if a mother has insufficient Vitamin D, it is probable that this will damage the development of the foetus' brain, resulting in reduced IQ, though this connection has not been directly confirmed.[15]

13—Edward Dutton, Salaheldin Farah, Attallah Bakhiet, Guy Madison, Yossry Ahmed, Sayed Essab, Mohammed Yahya, Mohammed Rajeh, "Sex Differences on Raven's Standard Progressive Matrices within Saudi Arabia and Across the Arab World," *Personality and Individual Differences* 134 (2018): 66-70.

14—Michael F. Holick, "Vitamin D Deficiency," *The New England Journal of Medicine* 357 (2007): 266-281.

15—Milou A. Pet and Elske M Brouwer-Brolsma, "The Impact of Maternal Vitamin D Status on Offspring Brain Development and Function: A Systematic Review," *Advances in Nutrition* 7 (2016): 665-678.

More broadly, *purdah* involves the intense control of females. If they dishonor their husband or their family, then the Prophet is quite clear that they have to be killed.

> *. . . the Messenger of Allah said: "Carry out the legal punishments on relatives and strangers, and do not let the fear of blame stop you from carrying out the command of Allah.*
> —Sunan Ibn Majah 3:20:254

It is plain from the above quote that it is acceptable to take the law into your own hands. And it is acceptable to kill a member of your family, such as your wife, if she does not obey you and dishonors you:

> *He sat before the Prophet (peace be upon him) and said: Apostle of Allah! I am her master; she used to abuse you and disparage you. I forbade her, but she did not stop, and I rebuked her, but she did not abandon her habit. I have two sons like pearls from her, and she was my companion. Last night she began to abuse and disparage you. So I took a dagger, put it on her belly and pressed it till I killed her. Thereupon the Prophet (peace be upon him) said: Oh be witness, no retaliation is payable for her blood.*
> —Sunan Abu Dawud 38:4348

Illicit sexual relations are noted to be a clear justification for honor killing:

> *It was narrated from Abu Hurairah that: The Messenger of Allah said: "No woman should arrange the marriage of another woman, and no*

> *woman should arrange her own marriage. The adulteress is the one who arranges her own marriage*
> —Sunan Ibn Majah 3:9:1882

Thus, Islam places females under a degree of stress that is less clear for comparable males. If they displease their father or their husband, then females may be killed with little consequence for the father or husband. There is evidence that stress reduces IQ, and in particular verbal IQ[16] and it certainly reduces IQ test performance[17] because, in essence, it means that people cannot "think straight." Insomuch as Islam helps to create suboptimal economic conditions, which we have already seen that it does, living under Islam would elevate stress levels for both men and women, reducing their ability to think coherently and thus reducing their intelligence.

On the plus side, honor killing can be regarded as the ultimate way of expressing your commitment to the group: You are so committed to the group that you are prepared to damage your own genetic interests, at least overtly, by, for example, killing your daughter or sister, if this is what the community believes is necessary. In terms of ethnocentrism, this kind of behavior has advantages on at least two levels. Firstly,

---

16—Philip Saigh, Anastasia Yasik, Richard Oberfield, Phill Halamandaris, and James Bremner. "The Intellectual Performance of Traumatized Shildren and Adolescents With or Without Posttraumatic Stress Disorder, *Journal of Abnormal Psychology,* 115 (2006): 332-340.

17—Joanna Moutafia, Adrian Furnhama, Ioannis Tsaousis, "Is the Relationship Between Intelligence and Trait Neuroticism Mediated by Test Anxiety? *Personality and Individual Differences* 40 (2006): 587-597.

the stain of dishonor is removed from the family; it is purged via the family killing its dishonorable member. Accordingly, the remaining members are once again sound marriage prospects or acceptable people with whom to do business, as they have signaled their religious commitment and, by extension, their ability to cooperate. It should be noted that in all societies, even non-religious societies, religiousness positively correlates with the two personality characteristics that are associated with cooperativeness; Conscientiousness and Agreeableness,[18] the essence of which is altruism. But the religious exhortation to kill dishonorable family members, much as it may create a stressful society for females, will tend to remove those who are not religiously committed, and who are thus uncooperative, from the gene pool before they are able to contribute to it. Accordingly, this aspect of *purdah* prevents the community from becoming gradually less ethnocentric. If you manifest particularly low religious devotion, and thus low ethnocentrism, then you will be killed.

The Effects of Polygamy

Polygamy is explicitly permitted in Islam, as we have already noted:

18—Jochen Gebauer, Wiebke Bleidorn, Samuel Gosling, Peter Rentfrow, Michael Lamb, Jeff Potter, "Cross-Cultural Variations in Big Five Relationships with Religiosity: A Sociocultural Motives Perspective," *Journal of Personality and Social Psychology* 107 (2014): 1064-1091.

> *And if you fear that you shall not be able to deal justly with the orphan girls, then marry (other) women of your choice, two or three, or four but if you fear that you shall not be able to deal justly (with them), then only one or (the captives and the slaves) that your right hands possess. That is nearer to prevent you from doing injustice.* —Sunan of Abu Dawood, 2128, Narrated Abu Huraira

The Prophet himself had multiple wives:

> *Whenever the Prophet intended to proceed on a journey, he used to draw lots amongst his wives and would take the one upon whom the lot fell. Once, before setting out for Jihad, he drew lots amongst us and the lot came to me; so I went with the Prophet; and that happened after the revelation of the Verse Hijab (i.e. veiling).* —Sahih Bukhari 7.157, Narrated Al Miswar bin Makhrama

However, certain Islamic nations have introduced restrictions. In Bangladesh, a man wishing to a take a second wife must make a formal application to a local council, which will hear opposing views on the matter, including those of his current wife and her family, and reach a decision.[19] In Pakistan, polygamy is only permissible with the written consent of the first wife, and it is increasingly socially unacceptable. Syria, Iraq, and Morocco have also placed restrictions on its practice.

19—Elora Shehabuddin, *Reshaping the Holy: Democracy, Development, and Muslim Women in Bangladesh* (New York: Columbia University Press, 2008), 90.

Tunisia has banned polygamy,[20] as has Turkey. Indeed, Turkey only recognizes civil marriages as valid.[21]

In comparison to monogamy, polygamy reflects a faster Life History Strategy; it is a movement towards quantity rather than quality. Polygamy involves the male distributing a (lower level of) investment over a larger number of females and, likely, a larger number of children. It is a means of hedging your bets. This tends to happen in an unstable ecology where you can be wiped out at any moment. In such an unpredictable environment, your one wife or your small number of children in whom you've invested a great deal could all suddenly die, meaning your investment had been wasted. So it makes sense to hedge your bets and maintain a harem of wives, investing relatively little energy in each. In comparison to a monogamous society, this will generally mean that fathers spend less time with their children, which may have some modest effect on intellectual development.

However, far more important is the kind of society that polygamy produces. Females are evolved to select for high status males; to marry "hypergamously"; meaning they "marry up." This is because females have more to lose from the sexual encounter than do males; because they can become pregnant. This makes them pickier than males and pickier with regard to specific traits. These females and their offspring are more likely to survive if the male is prepared to and able to support them, so it makes sense for females to be attracted, to a greater extent

20—Rubya Mehdi, *The Islamization of the Law in Pakistan* (London: Routledge, 2013), 162.

21—John Witte, *The Western Case for Monogamy Over Polygamy* (Cambridge: Cambridge University Press, 2015), 15.

than males are, to signs of high socioeconomic status or at least to signs of the ability to attain it.[22] The result of this, in Islamic countries, is that the wealthiest males will have multiple wives. Sunni teaching permits males to contract a maximum of only four wives. By contrast, Shia teaching permits men to take as many "temporary wives" as they can afford. In some Shia sects, it is also permitted to take concubines.[23] This means that the poorest males will be unable to find a wife at all. And, under Islamic Law, there are clearly very strict punishments, specifically death by stoning, for those who engage in extra-marital sex. The consequence of this, and it can be seen most obviously among polygamous hunter-gatherer tribes in which only about 40 percent of males ever have any children,[24] is a violent society in which, as in the animal kingdom, men fight over females. A further result is that groups of undesirable young males will gang together to look for females to gang-rape. As with lions, they will rape these females in an approximate pecking order and sperm competition will decide which, if any of them, passes on their genes, assuming the female does not have an illegal abortion. To a significant degree, Islamic Law obviates this problem by enforcing lethal punishment both for the rapist and the woman who is raped. The woman, if she had followed the Prophet's rules of *purdah*, would not have been raped, because she would have been

22—See David Buss, *The Evolution of Desire*, *op. cit.*; Edward Dutton, "Women Marry Up," In *Encyclopaedia of Evolutionary Psychological Science*, eds. Todd K.Shackelford and Viviana Weekes-Shackelford (New York: Springer, 2018).

23—Juliette Minces, *The House of Obedience: Women in Arab Society* (London: Palgrave Macmillan, 1982), 62.

24—See Lynn, *Dysgenics*, *op. cit.*

entirely covered, and thus not arousing to young males, and chaperoned by a male relative, who, it is assumed, not always correctly, would not harm her. So, a kind of anarchy is prevented by a combination of bloody punishments and the forcible regulation of female behavior. We can easily see how this would encourage the view that females should in very few respects be free, which would in turn demand the obedience of females. Thus, in an indirect sense, it would discourage critical thinking among females, this being an aspect of intelligence. It would also help to create a relatively stressful society and we have noted the negative effect that anxiety has on performance in intelligence tests.

But it also creates a further problem, of the kind that is present in prisons: a large number of extremely sexually frustrated young men with no access at all to females. And in cases where *purdah* is strictly observed, young men do not even get to *see* the faces of fertile females, other than those of close relatives. Research has indicated that pederasty is effectively socially acceptable in many Islamic countries, though homosexuality between two grown men is strongly taboo and severely punished.[25] In these countries, many males were sodomized as children, and they do the same to children when they themselves become adults. It has been found that pederasty is far more common in Islamic societies where females are heavily secluded than it is in those where they are less secluded. Indeed, it has been argued that part of the reason for the

25—See Wayne R. Dynes and Stephen Donaldson, *Asian Homosexuality* (London: Routledge, 1992); David F. Greenberg, *The Construction of Homosexuality* (Chicago: University of Chicago Press, 1990); and Michael Luongo, ed., *Gay Travels in the Muslim World* (London: Routledge, 2013).

widespread practice of pederasty in Ancient Greece was the seclusion of females.[26]

This theory is consistent with the widely documented phenomenon of males becoming temporarily homosexual if they are isolated from females for long periods of time, such as in prison, when they may become "prison gay,"[27] or at boys' boarding schools in England.[28] This presumably occurs because certain kinds of male, those who are not yet masculinized or are the least masculine, are the closest thing to females that is available or indeed visible, and people's sexual urges require fulfillment. Nineteenth and early twentieth century boys private boarding schools, known as "prep schools" and "public schools," were notorious hotbeds of pederasty. This would appear to be consistent with observations about Ancient Greece and with regard to the Islamic world. This occurred in a context in which females were absent or strongly controlled and secluded when they were not absent, and in a context where adult homosexuality was a serious crime.[29] Often this pederasty, at these kinds of schools, was sublimated into beatings that included a sexual element. Anthony Chenevix-Trench, whom we met earlier, was the, usually drunken, Eton headmaster in the 1960s. He would give ten pats to boys' bare buttocks before and after beating

26—Stephen O. Murray, "Some Nineteenth Century Reports of Islamic Homosexualities," in *Islamic Homosexualities: Culture, History, and Literature*, eds. Will Roscoe and Stephen O. Murray (New York: New York University Press, 1997).
27—Jeffrey Ian Ross and Stephen C. Richards, *Surviving Prison: Behind Bars* (New York: Penguin, 2002), 86.
28—Jonathan Gathorne-Hardy, *The Public School Phenomenon* (London: Harmondsworth, 1977).
29—*Ibid.*

them. During one such thrashing, he actually asked a boy if the boy thought that his headmaster was a "pederast." He would tell the boys that he loved them during such floggings. In 1964, Chenevix-Trench offered the future journalist Nick Fraser (who was 18) the choice between a bare bottom beating or expulsion. While patting Fraser's naked buttocks and whipping him, Chenevix-Trench "wept profusely" and told his victim how much he, the headmaster, "hated himself."[30]

As late as the 1920s, boys were sold off in Moroccan markets as sex-slaves.[31] Afghanistan, where women tend to completely cover themselves, is a notorious stronghold of pederasty. Men were sodomized themselves when they were children, and they expect to be able to do the same to others when they become adults. The boys are considered feminine and having sex with them, in effect, anally raping them, is euphemistically termed *bacha bazi:* "boy play."[32] Obviously, we would expect such acts to create trauma, and so reduce intelligence.

For strict Islamic societies homosexual statutory rape (in the Western sense) remains endemic. While pederasty is a terrible consequence of polygamy, polygamy itself can be argued to have positive effects, in terms of ethnocentric

---

30—Nicholas Fraser, *The Importance of Being Eton: Inside the World's Most Powerful School* (London: Short Books, 2012).

31—John R. Bradley, *Behind the Veil of Vice: The Business and Culture of Sex in the Middle East* (London: Palgrave Macmillan, 2010), 248.

32—Michelle Schut and Eva Van Baarle, "Dancing Boys and the Moral Dilemma of Military Missions," in *International Security and Peacebuilding: Africa, the Middle East, and Europe*, ed. Abu Bakarr Bah (Bloomington, IN: Indiana University Press, 2017).

behavior. In general, we would expect it to lead to a society with elevated levels of violence and distrust, due to the presence of such a high percentage of young, single men likely to be unable to find a wife nor have any hope of doing so. Islam deals with this problem through *purdah* and through very severe punishments for those who break the law. However, it has also developed the notion of *Jihad*; the idea that at certain points it is acceptable, indeed it is your duty, to fight others in the name of Islam and to lay down your life in the name of Islam. God's reward for dying in battle in monogamous societies, such as in Christian England, is the same as that accorded to any other Christian: to ultimately go to Heaven and sit beside God. In Islam, the enticement is far more sensual. All women will once again become young and beautiful in Heaven, even if they died elderly and unattractive:

> *Al-Hasan Al-Basri says that an old woman came to the messenger of Allah and made a request, O' Messenger of Allah make Dua that Allah grants me entrance into Jannah. The messenger of Allah replied, O' Mother, an old woman cannot enter Jannah. That woman started crying and began to leave. The messenger of Allah said, Say to the woman that one will not enter in a state of old age, but Allah will make all the women of Jannah young virgins. Allah Ta'aala says, Lo! We have created them a (new) creation and made them virgins, lovers, equal in age.*
> —Al-Tirmidhi, Jami` at-Tirmidhi (Surah Waaqi'ah, 35–37)

However, the reward for a male Muslim is far more specific; you go to Paradise where you will be attended to sexually for all eternity. The *Koran* describes what believing men will find in Paradise:

> *And endeared women.*
>
> *Indeed, We created them of a novel creation.*
>
> *And made all of them virgins.*
>
> *Loving their husbands and of equal age, fluent, and sweet of tongue.*
>
> *All these are for the people on the Right Hand.*
>
> *A multitude of those [on the Right Hand] shall be from former nations.*
>
> *And a multitude of those [on the Right Hand] shall be from the later generations.*
> —Koran, 56: 35-40

Muhammad is reported to have provided further detail, as quoted in many places in the *Hadith*.

> *Abu Umama narrated: "The Messenger of God said, 'Everyone that God admits into paradise will be married to 72 wives; two of them are houris (perfect female beings, existing only in paradise) and seventy of his inheritance of the [female] dwellers of hell (very beautiful women who were not Muslim in life).*[33] *All of them will have libidinous sex organs and he will have an ever-erect penis.'*
> —Sunan Ibn Majah, Zuhd, Book of Abstinence, 39

33—In order for the males to have all of these wives, non-Muslim women are transferred out of Hell to be part of their harems. See John Azumah, *My Neighbour's Faith: Islam Explained for African Christians* (Grand Rapids, MI: Hippo Books, 2009), 85.

> *Anas said, Allah be well-pleased with him: The Messenger of Allah said, upon him blessings and peace: "The servant in Paradise shall be married with seventy wives." Someone said, "Messenger of Allah, can he bear it?" He said: "He will be given strength for a hundred."* —Sifat al-Janna, al-`Uqayli in the Du`afa', and Musnad of Abu Bakr al-Bazzar

> *From Zayd ibn Arqam, Allah be well-pleased with him, when an incredulous Jew or Christian asked the Prophet, upon him blessings and peace, "Are you claiming that a man will eat and drink in Paradise?" He replied: "Yes, by the One in Whose hand is my soul, and each of them will be given the strength of a hundred men in his eating, drinking, coitus, and pleasure."* —Sifat al-Janna, al-`Uqayli in the Du`afa', and Musnad of Abu Bakr al-Bazzar

In other words, in return for your earthly sacrifice of laying down your life and, possibly, never having had your own harem and perhaps even dying a virgin (as the religious rules require of an unmarried man), you will receive a substantial and extremely desirable harem and a permanent erection with which to engage in sexual intercourse with them for eternity. Assuming this aspect of Islamic thought is genuinely believed, which it presumably is by those with sufficient fervor to sacrifice their lives for Islam, then this can be regarded as a strong motivator towards ethnocentric behavior.

Cousin Marriage

How can you create the cooperative society which wins the battle for group selection, when people are in an unstable ecology and so don't develop strong bonds? In addition to religion and patriarchy, a further strategy is cousin marriage. If everyone is closely related then males can be persuaded to invest in their families, and the broader society, because even if they've been cuckolded their wife's children are still relatively closely related to them. No Islamic text explicitly states that you *should* marry your cousin, but neither is it ruled out. The *Koran* states:

> *Prohibited to you (For marriage) are:- Your mothers, daughters, sisters; father's sisters, Mother's sisters; brother's daughters, sister's daughters; foster-mothers (Who gave you suck), foster-sisters; your wives' mothers; your step-daughters under your guardianship, born of your wives to whom you have gone in, no prohibition if you have not gone in; (Those who have been) wives of your sons proceeding from your loins; and two sisters in wedlock at one and the same time, except for what is past; for Allah is Oft-forgiving, Most Merciful.*
> —Koran, 4: 23

However, the Prophet Muhammad married Zaynab bint Jahsh, who was one of his nieces, as recorded in the *Koran*:

> *And [remember, O Muhammad], when you said to the one on whom Allah bestowed favour and you bestowed favour, "Keep your wife and fear Allah," while you concealed within yourself that which*

> *Allah is to disclose. And you feared the people, while Allah has more right that you fear Him. So when Zayd had no longer any need for her, We married her to you in order that there not be upon the believers any discomfort concerning the wives of their adopted sons when they no longer have need of them. And ever is the command of Allah accomplished.*
> — Koran 33:7

Muhammad also married his daughter Fatima off to his cousin Ali.[34] Accordingly, there is a sound case for arguing that cousin marriage is *Sunnah*; it is following the deeds of the Prophet and, as a Muslim, this is precisely what you should do.[35] Indeed, this is exactly what many Muslims do, overtly for this reason, and it is likely the reason why the more Muslim dominated areas of India have relatively low IQ. In the 1990s it was estimated that 61 percent of marriages in Pakistan were consanguineous ("cousin marriages") and similar levels of such marriages have been reported among Pakistanis in Norway. Around 33 percent of Pakistani marriages in Nottingham, in England's East Midlands, have been reported to be consanguineous while the prevalence of such marriages is also relatively high among Bangladeshis in the UK.[36] In the northern English city of Bradford in Yorkshire, which was roughly 27 percent South Asian in 2011, 63 percent

34—C.T.R. Hewer, *Understanding Islam: The First Ten Steps* (London: SCM Press, 2006), 123.
35—Alan H. Bittles, *Consanguinity in Context* (Cambridge: Cambridge University Press, 2012), 22.
36—Alan H. Bittles and Michael Black, "The Impact of Consanguinity on Neonatal and Infant Health, *Early Human Development* 86 (2010), 737-741.

of those of Pakistani descent are the products of first cousin marriage.[37] This inbreeding would have two effects. Firstly, it would increase the likelihood that Pakistanis and Bangladeshis would suffer from genetic disorders and thus decrease their average life expectancy and increase their level of infant and child mortality. This problem is made all the more prevalent by the fact that when two Pakistani cousins marry, it will usually be the case that each of their parents were cousins, and their grandparents and so on. Children of Pakistani descent in the UK accounted for about 3 percent of births in 2014, but 30 percent of those born with genetic disabilities.[38] It has been shown that a very significant reason (50 percent of cases) for relatively higher infant mortality among Pakistanis born in the UK, for example, is "a high rate of lethal malformations."[39]

The second effect of inbreeding is to reduce average intelligence,[40] something known as "inbreeding

37—Eamonn Sheridan, John Wright, Neil Small, et al., "Risk Factors for Congenital Anomaly in a Multiethnic Birth Cohort: An Analysis of the Born in Bradford Study," *The Lancet* 382 (2013): 1350-1359.
38— Steven Swinford, "First Cousin Marriages in Pakistani Communities Leading to 'Appalling' Disabilities Among Children, *Daily Telegraph,* July 7, 2015 https://www.telegraph.co.uk/news/health/children/11723308/First-cousin-marriages-in-Pakistani-communities-leading-to-appalling-disabilities-among-children.html (accessed August 15, 2019).
39—Sarah Bundey, Hasina Alam, Amritpal Kaur, Samina Mir, and Robert Lancashire, "Why Do UK-born Pakistani Babies Have High Perinatal and Neonatal Mortality Rates? *Paediatric and Perinatal Epidemiology* 5 (1991): 101-114.
40—Jan te Nijenhuis, Mart-Jan de Jong, Arne Evers, Henk van der Flier, "Are Cognitive Differences Between Immigrant and Majority Groups Diminishing? *European Journal of Personality* 18 (2004): 405-434.

depression." It is probable that the reason for this impact is double doses of deleterious mutations leading to a less well functioning nervous system, that would also be associated with poor general health. Low intelligence would further influence levels of child mortality, increase levels of poverty, and illness[41] and increase impulsiveness, as this is negatively associated with intelligence.[42] It has been found that, across 72 countries, national IQ and national rate of cousin marriage correlate at -0.62.[43] A study in India, comparing inbred and non-inbred families, found that inbreeding led to a genetically driven IQ decline of around 20 points on average, as well as more than double the rate of severe mental retardation.[44] Tacit justification of inbreeding is thus a key way in which Islam reduces IQ.

However, as with so many of the Prophet's IQ-reducing ideas, it is also going to increase ethnocentrism and my own study[45] has found that a higher rate of cousin marriage does indeed predict stronger ethnocentric attitudes and behavior. A key predictor of cooperation is genetic similarity, because the more genetically similar you are to somebody the greater is the degree to which

41—Lynn and Vanhanen, *Intelligence, op. cit.*
42—Noah A. Shamosh and Jeremy R. Gray, "Delay Discounting and Intelligence: A Meta-analysis, *Intelligence* 36 (2008): 289-305.
43—Michael A. Woodley, "Inbreeding Depression and IQ in 72 Countries," *Intelligence* 37 (2009): 268-276.
44—Mohd Fareed, Mohammad Afzal, "Estimating the Inbreeding Depression on Cognitive Behavior: A Population Based Study of Child Cohort," *PLOS ONE*. doi: 10.1371/journal.pone.0109585
45—Edward Dutton, Guy Madison, and Richard Lynn, "Demographic, Economic, and Genetic Factors Related to National Differences in Ethnocentric Attitudes, *Personality and Individual Differences* 101 (2016): 137-143.

cooperation with them assists you in indirectly passing on your genes.[46] This is why spinster aunties will spoil their nephews and nieces. They do not have their own offspring, so they pursue a strategy of kin selection, allowing them to indirectly pass on their genes, 25 percent of which are carried by each nephew or niece. There is a substantial body of evidence for "Genetic Similarity Theory," as set out by J. Philippe Rushton.[47] Couples are more genetically similar to each other than two random co-ethnics and they are more genetically similar on more heritable traits, something that is also true of same sex best friends. This is true of both physical and psychological traits; they are more similar on the more genetic dimensions of intelligence, such as *g*, than they are on the more environmentally influenced dimensions. Identical twins choose spouses and friends that are more genetically similar to their co-twin than do non-identical twins. Adoptions are more successful the more genetically similar the adoptive parents are to the adopted child. People mourn more for members of the family who resemble them to a greater extent and identical twins are far closer to their co-twin than are non-identical twins. Women prefer the bodily scents of males who are more genetically similar to themselves. People find a photograph of a face the most attractive when their own face is morphed into it, even though they do not recognize it as being their own face. People rate photographs of strangers as more trustworthy if their own face has been subtly morphed into these photographs.[48]

46—Frank Salter, *On Genetic Interests*, *op. cit.*

47—J. Philippe Rushton, "Ethnic Nationalism, Evolutionary Psychology and Genetic Similarity Theory, *Nations and Nationalism* 11 (2005): 489-507.

48—*Ibid.*; Dutton, "Women Marry Up," *op. cit.*

One study has shown that if you help a random co-ethnic, and you are from a country such as France, it is worth two percent of helping yourself, for percent of helping your child, eight percent of helping your uncle or nephew and so on.[49] But, in this context, helping a foreigner would be worth nothing, because two random co-ethnics will always be more related to each other than either will be to a foreigner, because ethnicities are essentially extended kinship groups.[50] Salter has set out the degree to which different racial groups are related to each other. Insomuch as different racial groups are genetic kinship groups, it will almost always be worth fighting any incursion of foreigners, because these will damage your genetic interests unless it is a very small number of foreigners and this damage is outweighed by some positive that the foreigners bring, such as a cure for some genetic disease that causes high mortality among your ethnicity. The more genetically dissimilar the two groups are then the more damage an increasingly smaller number of foreigner interlopers does to an individual member's genetic interests. Thus, if Australian Aboriginals were to invade areas held by the Bushmen, it would be worth Bushmen sacrificing their lives to repel even a small number of Australian Aboriginals, because the two groups are so genetically different. Cousin marriage helps to create a small and highly distinct gene pool. As such, it means that "foreigners" are particularly "foreign," elevating negative ethnocentrism. It also means that all co-ethnics are particularly closely related, elevating positive ethnocentrism.

49—Frank Salter and Henry Harpending, "J.P. Rushton's Theory of Ethnic Nepotism," *Personality and Individual Differences* 55 (2013): 256-260.

50—Salter, *On Genetic Interests, op. cit.*

Slavery

Slavery is unquestionably sanctioned in Islam. Most obviously, the females of defeated forces may be taken as sex slaves:

> *O Prophet! We have made lawful to thee thy wives to whom thou hast paid their dowers; and those (slaves) whom thy right hand possesses out of the prisoners of war whom Allah has assigned to thee*
> —Koran, 33: 50

> *. . .who abstain from sex, except with those joined to them in the marriage bond, or (the captives) whom their right hands possess . . .*
> —Koran, 23: -6

This will have an indirect impact on Muslim intelligence. Males operate a faster Life History Strategy than females because, especially in lawless societies, they have nothing to lose from the sexual encounter. If possible, they can have sex with numerous women, in the hope that some will become pregnant and produce surviving offspring, which is why males tend to be more promiscuous than females. Pursuing this strategy, it makes sense, if confronted with many females, to be more attracted to those who are more likely to produce surviving offspring. For this reason, males tend to be attracted to youth and beauty. Youthfulness is, obviously, a sign of fertility and thus the ability to become pregnant. The essence of beauty is a low number of mutant genes; mutant genes almost always being detrimental to health and, indeed, fertility. We tend to find symmetrical faces, and faces with specific ratio-distances between their different sections, attractive because they mean that a woman has been able to maintain a symmetrical phenotype in the face of disease and privation. Accordingly, she has an excellent immune system and thus a low level of mutation. In addition, she hasn't inherited many mutant genes as they relate to the face, and mutations tend to be comorbid. So males are far more interested than are females in youth and beauty. They are much less interested in traits that predict socioeconomic success, such as intelligence or impulse control.[51]

So, these rules within Islam positively encourage Muslim men to impregnate good-looking young women, without any thought for their psychological traits. In addition, impregnating them is especially justified if they are slaves. It tends to be those of relatively low intelligence who end up being taken as slaves, because, all else controlled for, the more intelligent group will tend to triumph, at least assuming equal levels of ethnocentrism. The more intelligent group will cooperate better, have superior weapons and it will pursue more successful battle strategies. Consequently, we would expect the enslaved females to be less intelligent than Muslim females. Even if high ethnocentrism Muslims have managed to overwhelm a group who are more intelligent than them, it is still likely to be that group's poorer (and so on average less intelligent) members who will be the most vulnerable and poorly protected and so the most likely to be taken as slaves. These females will be impregnated, producing children whose IQ is likely to be the average of the Muslim and conquered people's IQ, thus gradually helping to reduce Muslim average intelligence.

51—Buss, *The Evolution of Desire*, *op. cit.*

Battles are not the only means by which Muslims have enslaved infidels. Historically, when Barbary pirates used to raid the English and Irish coasts in the eighteenth century, it would be relatively poor people, unable to protect themselves, who would be taken captive and who would find themselves in the Moroccan slave markets.[52] Poverty correlates with low IQ, so this religious policy is likely to reduce Muslim IQ, though also ensure that its gene pool doesn't become damagingly small. It is recorded that the Moroccan emperor Moulay Ishmael the Bloodthirsty (1634-1727) fathered 888 children. These were produced by hundreds of concubines and 9 wives, including "Mrs Shaw, an Irish woman" who was taken as a slave by Barbary pirates during a raid on her native land.[53]

In modern times, this religious policy has seemingly manifested itself in so-called "Muslim Grooming Gangs" who seduce and then rape or abuse young, working-class native girls in poor, post-industrial towns, such as Rotherham in Lancashire.[54] Indeed, in 2018 it came to light that this had been happening in Oulu as well.[55] Clearly, any issue from such relationships would be likely to have relatively low IQ. More generally, it seems to be specifically low IQ Western women who marry Muslim males anyway, highlighting another way in which this

---

52—Giles Milton, *White Gold: The Extraordinary Story of Thomas Pellow and North Africa's One Million European Slaves* (London: Hodder and Stoughton, 2004).

53—Martin Daly and Margo Wilson, *Sex, Evolution and Behavior* (Boston: Willard Grant Press, 1983).

54—Jayne Senior, *Broken and Betrayed: The True Story of the Rotherham Abuse Scandal by the Woman Who Fought to Expose It.* (London: Pan Macmillan, 2016).

55—Dutton, *The Silent Rape Epidemic*, *op. cit.*

justification for Muslim out-breeding actually reduces IQ. It has been shown that just as females are attracted to males of high social status, they are also attracted to males from high status countries, because your nationality is an aspect of your status. A study of all multicultural marriages conducted in Finland in 2013, by myself and Guy Madison, proved that Finnish women are more likely to marry foreign males from countries that are wealthier than Finland. By contrast, Finnish males are more likely to marry foreign females from countries that are poorer than Finland. A noteworthy exception was Japan, possibly because Japanese females regard being White or European as being of high status.[56] I was motivated to test this simply by everyday observation. I was researching at Leiden University in 2003 when I realized that a lot of Chinese female postgraduates were particularly interested in going out with me. Other English men researching at the university remarked on having had a similar realization. In addition, I am married to a Finnish woman, and I live in Finland. Every Anglo-Finnish couple I have ever met in Finland is composed of a British male and a Finnish female. It appears to be similar when it comes to American-Finnish marriages or German-Finnish marriages. It is vanishingly rare to find this dynamic reversed. By contrast, when I have noted that Finnish men have foreign wives they will be, at least in the case of educated Finnish males, educated Chinese women whom they have met on business trips or women

56—Edward Dutton and Guy Madison, "Why Do Finnish Men Marry Thai Women But Finnish Women Marry British Men? Cross-National Marriages in a Modern Industrialized Society Exhibit Sex-Dimorphic Sexual Selection According to Primordial Selection Pressures, *Evolutionary Psychological Science* 3 (2017): 1-9.

from the wealthier ex-Communist countries: Estonia, Hungary or Slovenia. In the case of less educated Finnish males, they will be Thais or Russians.

However, within these Finnish multicultural marriages there was a clear social class divide. Where Finnish females married Muslim men, in other words married men from a poorer country, these Finnish females tended to be of low socioeconomic status[57] and also, albeit only in my own anecdotal experience, conspicuously not very good-looking. High status Finnish females married males from wealthier countries than Finland.[58] The explanation for this is relatively simple. At the group level, a slow Life History Strategy is positively associated with intelligence.[59] As such, working class females, who have relatively low intelligence as well as a fast LHS,[60] will be less concerned about acquiring a mate who is of high social status and has access to resources. In an unstable ecology, such traits are far less important than the ability to defend one's wife and children from sudden threats through extreme violence. Accordingly, they select for men who are physically tough, muscular and masculine-looking, and they tend not to care what country or ethnic background they are from. Also, faced with many such individuals to choose from, fast Life History strategists tend to be more attracted to genetic

57—Elli K. Heikkilä, "Multicultural Marriages and Their Dynamics in Finland," In *International Marriages in a Time of* Globalization, eds. Heikkilä and Brenda Yeoh, B (Hauppauge, NY: Nova Science Publishers, 2010).
58—*Ibid.*
59—Rushton, *Race, Evolution and Behavior*, *op. cit.*
60—*Ibid.*

dissimilarity. This is because an unstable ecology allows considerable genetic diversity. Basic needs are met, so organisms are not especially strongly adapted to the environment. This means that someone genetically very different is more likely to carry some beneficial adaptive gene that can be passed on to the offspring. Assuming the children of these unions are brought up Muslim then we would expect the Muslim IQ to be negatively impacted.

In terms of ethnocentrism, this form of breeding behavior is likely to simply ensure that Islam spreads further and further, meaning there are more and more people to fight for Islam, and for its genetic core. It also helps to create an important evolutionary balance. If Islamic genius levels are too low then this will be a problem, but if its ethnocentrism levels are too low this will also be a problem. An optimum level needs to be attained and breeding with slaves or infidels will have the potential to achieve that.

Islamic Civilization?

Of course, this begs the question of why, if Islam is so damaging to intelligence, it was able to develop a relatively advanced civilization, especially in its so-called "Golden Age." This was a period in which the Islamic world was divided into various caliphates between about the 800s and the sacking of Baghdad in 1258, after which time its leaders effectively declared that "the Gates of Learning were closed," and the Islamic world began to stagnate. By the sixteenth century, the Western world had caught up with the Islamic world, and had eclipsed it in terms of the standards of its civilization.[61]

During the reign of Harun Al Rashid (756-809) the "House of Wisdom" was established in Baghdad. Scholars were invited there from around the world and paid to translate Classical knowledge into Arabic, and to attempt to contribute to knowledge themselves. This period saw the introduction of a simpler writing system, of paper, of contributions to optics, mathematics and science, including al-Jahiz's (776-869) concept of the "struggle for existence" in zoology, and Nasir al-Din al-Tusi's (1201-1274) proto-evolutionary idea that humans were descended from animals. There were also contributions to medicine, such as the understanding that hospitals should be placed in areas where meat putrefies slowly. The Muslims invented an early sextant, a simpler system of numerals, and even the fork. This was all underpinned by a desire for knowledge, that was felt to be demanded by certain interpretations of the *Hadith*.[62] It is important, however, not to exaggerate the achievements of Islamic civilization. In comparison to those of Ancient Greece, they were minuscule. Islamic civilization was nowhere close to being as original and innovative as that of Ancient Greece. It merely rediscovered what the Greeks had done and made a few relatively minor developments on top of it. It was able to rediscover these things because it was so high in ethnocentrism that it could

61—Karen Armstrong, *The Battle for God: Fundamentalism in Judaism, Christianity and Islam* (London: HarperCollins, 2001).
62—Benson Bobrick, *The Caliph's Splendor: Islam and the West in the Golden Age of Baghdad* (New York: Simon & Schuster, 2012); Dutton and Woodley of Menie, *At Our Wits' End*, *op. cit.*

simply overwhelm vast areas of the Christian world that were, by then, in a Dark Age, but where the discoveries of the Ancient Greeks were preserved. American political scientist Charles Murray has documented the entire history of human achievement in his pithily entitled book *Human Accomplishment.*[63] In all areas, human achievement is absolutely overwhelmingly the product of the European and European-descended minds.

There are, however, a number of factors that would explain why Islam was able to develop an admirable civilization despite so many of its rules acting to reduce intelligence. Most obviously, by the 800s, a relatively large area was under Islamic rule, through conquest. This would have meant that even in spite of a propensity towards cousin marriage, many Islamic males would also have engaged in outbreeding, some of their wives being from very different areas. This would have led to a relatively large gene pool, including a significant influence from Europeans. At that point, much of North Africa and much of modern day Turkey and the Near East were populated by Europeans. In addition, the Muslims took Iberia, Crete, Cyprus, Sardinia, and Sicily. Persia, previously home to a major civilization, was also conquered. This large gene pool would have increased the likelihood of geniuses, whose innovations would have helped to develop Islamic civilization. Further, we would expect, based on Cold Winter's Theory, that the Europeans who mixed with the conquering Muslims would have had relatively high

63—Charles Murray, *Human Accomplishment: The Pursuit of Excellence in the Arts and Sciences, 800 B.C. to 1950* (New York: HarperCollins, 2003).

IQ in comparison to their conquerors, thus elevating the intelligence of the consequent Muslim population. Presumably this *Jihad*-induced intellectual boost was more than sufficient to overwhelm the negative consequences that various Islamic diktats would have on intelligence, thus permitting the development of Islamic civilization. A rational, tolerant form of Islam, known as *Falsafah,* became prominent.[64]

But, as with so many civilizations, the relatively luxurious conditions that this produced led to the upper class, at least, beginning to question traditional Islam. The religion became increasingly liberal, reflecting the low stress and high standards of living that had been created. There was the rise, for example, of *Sufi* Islam; a mysticism-based version of the faith that was also strongly focused around helping the poor. Helping the poor can be understood, to some extent, to mean helping those who have relatively low intelligence to survive and procreate. This ascetic, monastic movement would have enticed the highly intelligent who would be more educated and have more space to contemplate, just as it has been argued that the introduction of celibacy for Catholic priests likely reduced IQ in Europe.[65] Likewise, Sufism involved a highly ascetic movement and some of its followers, known as *Fakirs,* were world-renouncing celibates.[66] This kind of movement would have been attractive to high-IQ men, due to the association between intelligence and Openness-Intellect.

64—Graham Oppy, *Atheism: The Basics* (London: Routledge, 2018).
65—Lynn, *Dysgenics, op. cit.*
66—Green, *Sufism, op cit.*

Islam prohibited both infanticide and, other than very early in the pregnancy, abortion. German biochemist Gerhard Meisenberg has noted that we would expect both of these, due to their repulsiveness or dangerousness, to be done out of sheer desperation and thus involve women of low intelligence. Thus, Islam compelled women of low intelligence to keep their children. Also, Islamic advances in medicine, and an historical lack of prohibition of *coitus interruptus*, meant that relatively reliable contraception was developed in the Islamic world and was specifically sanctioned by Islamic authorities in the ninth century.[67] Contraception was not developed in the Western world until the late nineteenth century. Even though Islam condoned polygamy, we would expect that, as in our own time as we will see below, the more intelligent, in a context in which wealth was associated with lower child mortality, would be better at using contraception and would be more likely to only want a small family, especially under conditions of relatively low mortality salience. They would also be less fervently religious and less motivated by religious injunctions to breed. Members of the higher classes would also be aware that the likelihood of one of their children not reaching adulthood was relatively low, where poorer (on average less intelligent) people would, by contrast, have far higher mortality salience. Accordingly, the less intelligent would ultimately outbreed them and the "Gates of Learning" would close.

A view of Islam that was far less intellectual would thus arise. In many ways, this reactionary form of Islam

67—Şevket Pamuk and Maya Shatzmiller, "Plagues, Wages, and Economic Change in the Islamic Middle East, 700-1500," *Journal of Economic History* 74 (2014): 196-229, 216.

was spearheaded by the revival based around Persian theologian Abu Hamid Al-Ghazali (c.1058-1111). A kind of proto-fundamentalist, Al-Ghazali was strongly critical of science and espoused a return to an idealized past, which he associated with the religious zeal of the early Islamic community.[68] To make matters worse, Mongol invasions of the Islamic world, between 1206 and 1405, would have created chaos, elevated stress, and so pushed people even further towards anti-rational religiousness, in particular in a context where the predominant form of Islam was, at that time, relatively liberal.[69] This decline in intelligence, and related conservative revival, would explain why 64 percent of important Muslim scientists in *The Encyclopedia of Muslim Scientific Pioneers* lived before 1250 and almost 100 percent lived before 1750.[70] If we restrict the range to those who died between 700 and 1699, then 50/195 of the Islamic scientists on the current Wikipedia article "List of Muslim Scientists" died in the 1000s, the largest group by century of death.[71] This is in a context in which there was no significant population rise. So this is the peak of Islamic innovation. Interestingly, this is also the approximate centre point of Islam's "Golden Age" and in the following centuries there was a decline in innovation, until 1258, when the Golden Age is considered to be over.

68—Kenneth Garden, *The First Islamic Reviver: Abu Hamid Al-Ghazali and His Revival of the Religious Sciences*, (Oxford: Oxford University Press, 2014).
69—See Peter Jackson, *The Mongols and the Islamic World: From Conquest to Conversion* (New Haven, CT: Yale University Press, 2017).
70—Gerhard Meisenberg, *In God's Image: The Natural History of Intelligence and Ethics.* (Kibworth, Leicestershire: Book Guild Publishing, 2007), 85.
71—Dutton and Woodley of Menie, *At Our Wits' End*, *op. cit.*

Standards of living had fallen, based on the purchasing power of a family of four in Baghdad and Cairo, by 250 percent between approximately 800 and 1200. They rose by 57 percent between 1200 and 1600, in tandem with the strongly religious Ottoman Empire, and then declined once more.[72] This long-term decline in wealth is what we would predict if Islamic society were becoming less intelligent, national wealth being a robust correlate of national IQ.[73]

Thus, Islam reached a high point of civilization by a combination of conquering civilizations languishing in a Dark Age that had preserved the knowledge from their Golden Age, interbreeding with subjugated peoples of relatively high average intelligence, and by effectively moving away from the traditional teachings of Islam. The Renaissance is often regarded as a turning point in the history of Western civilization. It is the time when the West rediscovered the knowledge of the Ancient Greeks and began to seriously build upon it. By this time, however, Islamic civilization had pretty much stagnated and was entering a long period in which the original form of Islam would be followed: a long period of becoming, in comparison to the West, less intelligent and more ethnocentric. For the reasons we will outline in the next chapter, there would have been, under pre-industrial conditions, selection for intelligence in the Islamic world. More intelligent men would have become wealthier, leading to their having more wives, more offspring, and, due to superior living conditions and nutrition, more

72—Pamuk and Shatzmiller, "Plagues, Wages, and Economic Change in the Islamic Middle East, 700-1500," *op. cit.* Table 1.
73—Lynn and Vanhanen, *Intelligence*, *op. cit.*

offspring that survived childhood. But so many other factors, those we have already discussed, would have suppressed intelligence in the Islamic world, meaning that it would have at best risen much slower than in the West and, very probably, stagnated or declined.

Does Islam Make You Intelligent?

It should be noted that, for some people, aspects of Islam may actually make them *more* intelligent, particularly if their IQ is very low and they live in a non-Islamic society. Most obviously, committed Muslims will generally go to Friday prayers and follow assorted other Islamic dictates, such as not drinking alcohol or not eating pork. The ability to do these things, at least in a non-Islamic society, involves some degree of self-control, restraint, planning and future-orientation. These characteristics, as we have discussed, are all weakly associated with intelligence. It would follow that if a person who was of very low intelligence were to become Muslim then he would begin associating with people who were more intelligent than himself. As psychologist James Flynn has shown, a key dimension to the environmental component of IQ is whom you spend your time with. If you spend your time with people who are more intelligent than you are, then this is likely to push your IQ to its phenotypic upper limit, to the highest level that it is genetically possible for it to be, and thus to make you more intelligent.[74] This phenomenon is the reason why the heritability of IQ is

74—Flynn, *Does Your Family Make You Smarter?*, *op. cit.*

quite low in childhood, but rises to 0.8 in adulthood. As adults, people create their own environment, reflecting their own innate intelligence, no longer being shackled to the environment that reflects their parents' innate intelligence, which may be higher or lower than theirs.

Similarly, it might be averred that to have a consistent worldview, even one based around religious dogmas, requires a certain level of intelligence. Meisenberg and his colleagues[75] have brought together data worldwide on the relationship between educational level and religiousness. In most of the world, education level is weakly negatively associated with belief in God and strength of belief in God. However, this is not the case in Sub-Saharan Africa. In Sub-Saharan Africa, religiousness is usually positively associated with education level and, by extension, intelligence. A probable reason for this is that the surveys upon which Meisenberg and his colleagues draw focus on "belief in God" and on world religions. However, in many parts of Africa, those of low socioeconomic status will tend to still adhere to tribal religions based around the worship of spirits, meaning that they will be marked as not being "religious" or believers in "God." The more, intelligent, in this context, will be attracted to Christianity or Islam because, as Meisenberg and his colleagues summarize:

> *In the least developed world region, sub-Saharan Africa, this relationship is more often positive than negative. Although unusual in today's world, this*

75—Gerhard Meisenberg, Heiner Rindermann, Hardik Patel, Michael A. Woodley, "Is It Smart to Believe in God? The Relationship of Religiosity With Education and Intelligence, *Temas em Psicologia* 20 (2012): 101-120.

> *result confirms the anthropological observation that elaborated religion plays a minor role in many simple, small-scale societies. Religious belief and ritual both increase with rising cultural complexity . . . Indeed, before the birth of modern science, the major world religions offered the most sophisticated explanations of the human condition and of natural, psychological and social phenomena in general.*[76]

In other words, among those with very low intelligence, who are simply insufficiently bright to be able to comprehend a purely scientific worldview, some form of religiousness is the next best option in terms of the most sophisticated and parsimonious worldview. As Danish psychologist Helmuth Nyborg has noted, people will tend to become distressed if the world does not make sense to them.[77] For those whose IQ is below a certain level, when controlling for personality factors, as personality also influences the kind of worldview which is found attractive,[78] a purely scientific worldview is not satisfying, because there are significant aspects of that worldview which they simply cannot understand. Accordingly, a religious worldview, usually accepting many aspects of science, gives them the satisfaction which they require. For people with even lower intelligence, such as very young children, even this is beyond them, so they see the world through fairy tale beliefs in which spirits or non-physical beings appear to be behind every aspect of the

76—*Ibid.*, 115.

77—Dutton, Nijenhuis, Metzen, Linden, and Madison, "The Myth of the Stupid Believer, *op. cit.*

78—For example, as we have discussed, people who are high in empathy tend to believe in God, while people who are low in it, who are high on the autism spectrum, tend not to.

world. This is very similar to the worldview held by many hunter-gatherer societies.

This summary is in line with American philosopher Nathan Cofnas' argument that monotheistic religiousness was, under Darwinian conditions, selected for at the same time as intelligence: they co-evolved together.[79] Universalist religions are very different from tribal ones. Membership of a monotheistic religion is via religious belief and practice rather than simply via being a member of a particular ethnic group. Intelligence, avers Cofnas, predicts the ability and desire to cooperate, the ability to trust others, and the ability to innovate. These factors would mean that the cleverer populations would develop into larger polities with higher and higher levels of genetic diversity: they would be the groups who would build cities where constant fraternization with non-kin would become a significant issue. Accordingly, it would make sense that these kinds of societies would increasingly adopt a belief in a moral god. This moral god would force people to be altruistic even to strangers. It would also follow that such a society would innovate a universal form of religion, because its members would increasingly be from diverse kinship and ethnic groups and the society would be continuously expanding into new ones. Adherence to a universalist religion would, maintains Cofnas, become the fundamental marker that you were part of the "in-group"; that you could be trusted, because you believed in the same (moral) god who was watching over you and influencing your actions.

79—Nathan Cofnas, *Reptiles With a Conscience: The Co-Evolution of Religious and Moral Doctrine* (London: Ulster Institute for Social Research, 2012).

This all implies that for some people, especially those of very low intelligence, Islam will involve associating with people who are, on average, more intelligent than then. And Islam will push them to think in a way that is more sophisticated that the way in which they are used to thinking but, nevertheless, a way in which they have the ability to think, given the right circumstances and interlocutors. However, such people will lack the ability to think in a more scientific fashion, meaning that if they are only exposed to that then they are more likely to simply "drop out" into associating with people of their own intelligence level and, accordingly, not reaching the phenotypic maximum limit of their IQ. For people of very low intelligence living in Islamic societies, there will be all of the negative influences Islam has on intelligence to contend with. This, however, is not so clear in Western countries. In such societies, it may well be that Islam elevates both the IQ and the ethnocentrism of those who have relatively low intelligence.

An example of this can be see, I suspect, in Chicago. The average IQ of an African American is 85.[80] In June 2019, I visited the Chicago district of Englewood, which is 90 percent Black.[81] It was the middle of the day, as my guide, pistol at the ready, refused to go anywhere near the area after dusk, for fear of carjacking. But, even so, the evidence that we were in a low IQ area was everywhere:

80—Lynn, *Race Differences in Intelligence*, *op. cit.*

81—Manny Ramos, "As Black Population Drops, Hispanics are Drawn to Greater Englewood, *Chicago Sun-Times*, March 10, 2019, https://chicago.suntimes.com/2019/3/10/18313833/as-black-population-drops-hispanics-are-drawn-to-greater-englewood (accessed August 15, 2019).

pot holes in the road, uncut grass by the pavements, dilapidated houses that were obviously occupied, bars across windows and doors (implying a high crime rate), abundant fast food outlets and shops exclusively selling alcohol, litter in the streets, graffiti, and lots of people simply "hanging around" with nothing to do. Things got even worse as we turned off the main road. Not only were there more and more dilapidated houses, but it became clear that crime, in particular drug gangs, was such a problem that the community-minded Englewood residents had established a system so that children could get to and from school without being accosted by dealers. There were bright signs indicating a specific safe path, which they were to follow, which would seemingly be guarded by local volunteers. This is what to expect of an area with very low IQ. A few miles down the road, on Stony Island Avenue, was the impressive Mosque Maryam, a converted Orthodox church, the home of the Nation of Islam, an almost exclusively Black Islamic sect. The Black men in the car park were suspicious of us at first, but when another (more senior) member decided to be friendly this suspicion soon melted into congeniality. It was quite clear, not least from the testimonies of those to whom I spoke, that this group seemed to rescue young Blacks from a life of crime and give them a clear, structured and meaningful path. In doing so, in many cases, it would likely have elevated their IQ to its phenotypic limit. So, this positive intellectual aspect to non-liberal Islam must be borne in mind. But, as discussed, overall and for most Western people, the intellectual effect is going to be negative. Indeed, it might be argued that were non-religious Western people with low IQs to embrace many forms of Christianity they would receive all of the intellectual benefits of Islam without any of the inherent problems such as sleep disturbance and the need to fast.

Chapter 6

Why is the West in Decline?

Too Clever by Half

We have now explored the many ways in which Islam reduces intelligence but, in so doing, increases ethnocentrism. In this chapter, we will explore why, at the time of writing, this is causing Islam to spread while the West is left in retreat. The key reason is that intelligence is only a good thing, in evolutionary terms, up to a point. Beyond that point, intelligence is maladaptive, to the extent of causing the highly intelligent group to be wiped out. And Islam, because it "makes you stupid," ensures that that point is not reached

Until the American sitcom, *The Big Bang Theory* evolved into a predictable replacement for *Friends* I thought it was rather perceptive and funny. And it also presented, and sent up, a crucial nuance with regard to the importance of intelligence. Intelligence predicts success in all aspects of life whether at the individual or group level, but it is not an absolute predictor of success and sometimes it's only a relatively weak predictor of success. Other factors, such as personality traits or even physical appearance, are sometimes far more important. *The Big Bang Theory* presents this dimension to life. Penny is, apparently, not especially bright, to the extent that, at one point, she starts doing a History degree not only at a "community college" but specifically at a community college with a slogan: "a place of learning, a place of knowledge."[82] Her scientist friends are so concerned that she will flunk the course that they secretly alter her, extremely bad, essay, so that she can get a decent mark. Penny, however, has no problems dealing with people. Towards the end of the show's run, she manages to get a job as a saleswoman. She turns out to be very good at at it and it gets to a point where her salary is considerably higher than that of her physicist husband Leonard. Penny is warm-hearted, gregarious, practical, and socially skilled. She is high in what is sometimes termed "emotional intelligence"; in theory of mind. However, research indicates that this only weakly correlates with general intelligence, at around 0.3.[83] Intelligence *does* predict the ability to solve social problems, probably because it allows you to better solve any problem through applying previous knowledge to it, something called "crystallized intelligence." Intelligent people will have learnt, through prior observation, how other people's minds work, allowing them to empathize and attempt to solve social problems on this basis. However, a much better predictor of getting on with people is simply the person's Agreeableness and, specifically, "empathy" which is a key

82— *The Big Bang Theory*, "The Extraction Obliteration," November 1, 2012, Season 6, Episode 6, *CBS*.
83—Scott Barry Kaufman, Colin DeYoung, Diedre Reiss, and Jeremy Gray, "General Intelligence Predicts Reasoning Ability for Evolutionarily Familiar Content, *Intelligence* 39 (2011): 311-322.

trait of Agreeableness. People who score high on this trait have more friends; they know how to deal with people.[84]

Sheldon, in particular, is, again at the beginning of the show's run, very different from Penny. He has extremely high general intelligence. But one of the downsides of this is that the higher a person's IQ is, the weaker is the connection between the different kinds of intelligence. Thus, Penny has about average intelligence and she is about averagely intelligent in all of the different kinds of intelligence, such as verbal, spatial and linguistic. Sheldon has an astronomically high IQ and, like many people with such a profile, it is driven by his having quite fantastically high abilities on *one* particular measure of intelligence;[85] in his case Mathematics. The result of this is something that English psychologist Bruce Charlton has termed the phenomenon of the "clever silly." People with very high IQ possess a kind of skewed intelligence. The relationship between the different kinds of intelligence becomes weaker the more intelligent someone is.[86] This means that extremely intelligent people actually find some relatively-weakly *g*-loaded tasks extremely difficult, their intelligence being so strongly a matter of *g*. Consistent with this, Sheldon cannot drive a car, deal with money, or deal with people. He is cold, blunt and utterly lacking in tact or social skill. When Sheldon

84—See Nettle, *Personality*, *op. cit.*
85—Michael A. Woodley, "The Cognitive Differentiation-Integration Effort Hypothesis: A Synthesis Between the Fitness Indicator and Life History Models of Human Intelligence, *Review of General* Psychology 15 (2011): 228-245.
86—*Ibid.*

has to rescue Penny from the bath and her boyfriend, Leonard, is upset that Sheldon may have seen Penny "naked," Sheldon takes this completely literally and responds, "Oh, relax! It was just her bottom and her breasts."[87] In addition, intelligence correlates with certain personality-type traits, likely because having these improves your ability to solve problems. Intelligence correlates at about 0.3 with Openness-Intellect,[88] which involves being open-minded to unusual possibilities and fascinated by intellectual, abstract issues. In extreme cases, this means disinterest in, and even an inability to deal with, practical issues. Sheldon, for example, allows his pay checks to pile up in a drawer, unopened, because, "Most of the things I'm planning to buy haven't been invented yet."[89]

There is also evidence of autism traits, such as lack of social skill, hypersensitivity to stimuli, rigidity and being easily overwhelmed, at outstanding IQ levels.[90] Sheldon manifests all of these symptoms. This makes sense because high functioning autism involves being obsessed with rules and systems,[91] and this will often help you to solve cognitive problems. It may also help to explain the negative association between religiousness and intelligence, because autism is robustly associated with

---

87—*The Big Bang Theory*, "The Werewolf Transformation," Season 5, Episode 18, February 23, 2012.
88—Nettle, *Personality*, *op. cit.*
89—*The Big Bang Theory*, "The Focus Attenuation," Season 8, Episode 5, October 13, 2014.
90—Ruth Karpinski, Audrey Kinase Kolb, Nicole Tetreault, Thomas Borowski, "High Intelligence: A Risk Factor for Psychological and Physiological Overexcitabilities," *Intelligence* 66 (2018): 8-23.
91—Baron-Cohen, "The Extreme Male Brain Theory of Autism, *op. cit.*

low levels of religiousness.[92] Autism is characterized by an inability to discern mental states from physical cues and a cold, systematic approach to the world. It would follow that autistics would not conceive of the world as related to a kind of mind, with the world providing evidence of the workings of this mind. In addition, again in the earlier seasons, Sheldon is obsessively focused on his work and has no interest at all in females, let alone in having children. His level of Intellect, which as we have seen crosses over with intelligence, is so high that he is simply not interested in such mundane things. Perhaps you know someone like Sheldon Cooper—a man or woman with little interest in sex or procreation but possessed by a passion for science, idealism, politics, or even a hobby. Such eccentrics may have forgone immediate Darwinian imperatives, but charitably could be said to operate at the level of the group.

The Collapse of Darwinian Selection on Intelligence

In other words, there is a definite downside, in terms of the evolutionary imperative to pass on your genes, to being extremely intelligent and, especially, to having outlier high intelligence. It can make you socially blind,

92—Catherine Caldwell-Harris, Caitlin Fox Murphy, Tessa Velazquez, Patrick McNamara, "Religious Belief Systems of Persons with High Functioning Autism," Proceedings of the 33rd Annual Meeting of the Cognitive Science Society, Austin, Tx.: Cognitive Science Society, 3362-3366, http://csjarchive.cogsci.rpi.edu/proceedings/2011/papers/0782/paper0782.pdf (accessed August 15, 2019).

unattractive to the opposite sex, uninterested in the opposite sex, and thus unable to have children. We know, however, that under pre-industrial conditions, intelligence was *positively* associated with completed fertility: with how many of your children survived into adulthood and outlived you. There was a robust positive association between wealth and intelligence and wealth predicted completed fertility. All of the evidence for this has been set out in the book, *At Our Wits' End: Why We're Becoming Less Intelligent and What It Means for the Future,* by myself and Michael Woodley of Menie.[93]

We know from wills and parish records that in Early Modern Europe, the richer 50 percent of the population had about double the number of surviving children compared to the poorer 50 percent of the population. This was because wealthier parents lived in better conditions and had better access to food and specifically to more nutritious food. Thus, while the average child mortality rate was about 40 percent in pre-industrial England, it was much lower among the wealthier and much higher among the poorer, many of whom would find that all their children simply died young. The result of this was that average IQ increased every generation, as those of low IQ, occupying the lowest rung of the socioeconomic ladder, died out every generation. This created a society of social descent in which people had to move down the social ladder to fill the spaces left open by the death of the previous generation's lowest rung. Accordingly, most Early Modern males occupied a lower rank than that of the father: the sons of gentlemen became yeomen (doing some farm work themselves), the sons of yeomen became

93—Dutton and Woodley of Menie, *At Our Wits' End, op. cit.*

husbandmen (taking to the plough themselves), the sons of husbandmen mixed a smallholding with doing labor for husbandmen or yeoman, and so on.[94]

Consequently, intelligence increased every generation. This is clear because there is evidence that the population prevalence of genes associated with high intelligence increased across time. The per capita number of geniuses and extremely important inventions increased across time. And other proxies for intelligence increased between the Middle Ages and about 1800, despite very little change in living standards: literacy and numeracy rose, interest rates (a marker of future orientation) fell, people became less cruel to animals and to other people, corruption levels fell, and political stability and democracy increased, among other indicators.

The result of this was that our IQ was so high by about 1800 that the IQ of our most intelligent people, our so-called "smart fraction,"[95] was astronomical. Accordingly, they made incredible scientific breakthroughs, resulting in the Industrial Revolution. The resultant increased and more efficient production led to a massive rise in living standards and also in public health. Food and clothes became considerably cheaper, a huge food surplus could be produced, and sanitation was enormously improved. People could more easily heat their houses. Medicine developed and children could be inoculated against formerly killer diseases. Consequently, the population spiked, as people continued to have large numbers of

94—See Gregory Clark, *A Farewell to Alms: A Brief Economic History of the World* (Princeton: Princeton University Press, 2007).
95—See Rindermann, *Cognitive Capitalism, op cit.*

children, in the hope that some would survive, but, inoculated and living in far healthier conditions, they mostly did survive. Until the Industrial Revolution, there was a positive correlation between wealth and completed fertility, due to high child mortality among the poor. By the end of the nineteenth century, there was no such correlation. Indeed, the correlation eventually became negative: poverty and thus *low* intelligence was what predicted completed fertility.

How did this dramatic change come about? Reliable contraception was developed in the late nineteenth century and it was taken up by the more educated; those of higher intelligence. Being of higher intelligence, they had the ability to understand how to use contraception, the impulse control to actually use it, and the desire to use it. Being highly intelligent, they understood that their offspring were all likely to outlive them and they'd be better off financially with a smaller family, investing more attention and money in each child so that these children could duly succeed. Consequently, large families soon became something you had by accident because you were too impulsive to use contraception, in other words too low in intelligence to use contraception. In Western countries, in modern times, the correlation between IQ and number of children among middle aged people is about -0.2. Low intelligence now predicts fertility. And this correlation is stronger among women than men.

The Industrial Revolution has led to such a relaxation of Darwinian harshness that our stress levels are particularly low in comparison to our IQ. Consequently, towards the end of the nineteenth century people started to become decreasingly religious, questioning not just the existence of

God but, as a consequence, all of the likely evolutionarily adaptive traditions that were upheld by the state religion. The result was the rise of ideologies that strongly questioned Western traditions, one of which was Feminism. This resulted in equality for women which, in turn, resulted in more and more women going into higher education and taking up skilled jobs.[96] This has meant that, at the time of writing, more intelligent women spend most of their 20s and even the first half of their 30s focused on their careers. They have children late and they have a small number of them, if any at all. The most intelligent women, it should be noted, simply don't have children, perhaps because they are very high in Intellect and so more interested in other things. Also, women are increasingly more educated than men and have the capacity to earn just as much.[97] In that females are evolved to marry hypergamously in terms of status, there is, as a consequence, a dearth of attractive males if you a highly intelligent female, an issue explored by American researcher, F. Roger Devlin.[98] By contrast, the least intelligent females drop out of school, become mothers, beginning in their teens, to multiple children by multiple men. They likely become grandmothers by the time their more intelligent peers become mothers. This means that females of low intelligence not only have more children but more generations as well. Thus, it is entirely understandable that the negative intelligence-fertility nexus should be stronger among females.

96—See Joanna Williams, *Women vs Feminism: Why We All Need Liberating from the Gender Wars* (Bingley, West Yorkshire: Emerald Publishing, 2017).
97—*Ibid.*
98—F. Roger Devlin, *Sexual Utopia in Power: The Feminist Revolt Against Civilization* (San Francisco: Counter Currents Publishing, 2015).

To make matters worse, the collapse of religion has led to ideologies that preach not just sexual equality but social equality. Consequently, we have an extremely generous welfare system in many Western countries. In the 1950s, being the single mother of an illegitimate child was, partly for religious reasons, very much unacceptable, and such children would often be put up for adoption. English psychologist Adam Perkins has charted how attitudes have changed in his book *The Welfare Trait.*[99] Beginning in the 1970s, illegitimacy became ever more acceptable and welfare payments became ever more generous, more than enough for the mother to happily live off, squandering the lavish "child benefit" received for each child on alcohol and other luxuries, while feeding the children a poor diet. Perkins has shown that in the UK, if we compare households where both parents are working, households where one parent is working, and households where neither parent works (meaning both are on welfare), only the welfare households have above replacement fertility. Being on welfare is predicted by very low IQ.[100] And, of course, the ideology of Multiculturalism has led to mass immigration from low IQ, non-Western countries, further reducing intelligence.

As Woodley of Menie and myself show in detail in *At Our Wits' End,* it is quite clear that Western intelligence is declining and it is declining for genetic reasons, as the above causes would predict. Reaction times increased between 1880 and the year 2000 commensurate with an IQ drop of 15 points. We are getting worse at a variety of robust correlates of intelligence: color discrimination

99—Perkins, *The Welfare Trait, op. cit.*

100—Herrnstein and Murray, *The Bell Curve, op. cit.*

(if you can discriminate better between properties you can better solve problems), creativity, use of technically complex words, the ability to count backwards, per-capita levels of genius and major innovation, and simply on IQ tests. The latter decline had long been covered up by our enormous rise in a weakly *g*-loaded and highly environmentally sensitive ability, analytic thinking, cloaking the decline on IQ tests and resulting in the imperfect IQ tests showing a secular *rise* in scores across the twentieth century. But now that we have reached our phenotypic limit in this ability, the underlying intelligence *decline,* which is on the more *genetic* aspects of intelligence, has begun to reveal itself in Western countries, since about 1997. Most importantly, research from Iceland has shown that across three generations the population frequency of alleles associated with very high education levels, and thus high intelligence, has decreased.[101] So, Western people are becoming less intelligent and, in comparison to Muslim immigrants, they are also failing to breed. This failure is predicted by their higher levels of intelligence and also by their lower levels of religiousness, because religiousness predicts fertility, even among Western people and when controlling for intelligence, SES and other such confounds.[102]

101—Augustine Kong, Michael Frigge, Gudmar Thorleifsson, et al., "Selection Against Variants in the Genome Associated With Educational Attainment," *Proceedings of the National Academy of Sciences*, January 31, 2017, 114: E727-E732, https://www.pnas.org/content/114/5/E727 (accessed August 15, 2019).

102—See Lee Ellis, Anthony Hoskin, Edward Dutton and Helmuth Nyborg, "The Future of Secularism: a Biologically Informed Theory Supplemented with Cross-Cultural Evidence *Evolutionary Psychological Science* 3 (2017): 224-242.

The Collapse of Darwinian Selection on Everything Else as Well

Why does religiousness predict fertility and, for that matter, health, with which it correlates at 0.3?[103] Having originally trained in Religious Studies, I used to assume that it was all a matter of environment. As American psychiatrist Harold Koenig argues, belief in God reassures you that someone is watching over you and will make sure that everything will be okay in the end. It tends to mean that you are part of a religious community of friendly people whom you regularly meet. This reduces psychological stress which, in turn, reduces physical stress, which means that you are less likely to become physically or mentally ill.[104] In addition, because we are evolved to be religious, religions tend to present adaptive behavior as religious imperatives. It is God's will that you don't drink too much alcohol, engage in promiscuous sex, eat animals the hunting of which might get you seriously injured, eat animals likely to contain parasites, and so on. Thus, for that reason as well, religious people are healthier, they have lower morbidity, and if they get seriously ill then they are more likely to recover than are non-religious people.

The problem with this explanation is that it fails to explain certain religion-health associations that are nothing to do with the environment. For example, one study found that church-goers have better immune systems than those

103—Harold Koenig, "Religion, Spirituality, and Health: The Research and Clinical Implications," *ISRN Psychiatry* (2012), http://dx.doi.org/10.5402/2012/278730 (accessed August 15, 2019).
104—*Ibid.*

who do not go to church, even when you control for all manner of environmental factors that might explain this.[105] Another study found that church-goers were more likely than non-church-goers to carry alleles that are negatively associated with suffering from depression.[106] This implies that the documented negative association between suffering from depression and religiousness is not causal, or is not simply causal. Religiousness does not simply reduce the extent to which you suffer from depression. Rather, religiousness and not suffering from depression are underpinned by the same thing.

Now, having studied Religious Studies, initially as part of a Theology degree, I was only ever told about the environmental reasons for differences in levels of religiousness. Environmental and cultural determinism was the foundation of my degree and my subsequent doctorate, as it is in most Humanities subjects. When I realized that there was a significant genetic component to religiousness, I accordingly felt rather betrayed by my Religious Studies Professors. And this left me somewhat skeptical of purely environmental explanations. It occurred to me that the religion-health nexus could actually be mainly genetic. As we have seen, we are evolved to be religious and, indeed, to a very specific form of religiousness: the collective worship of a moral god. This is the kind of religiousness that was prevalent

105—Harold Koenig, Harvey J. Cohen, Linda George, et al., Attendance at Religious Services, Interleukin-6, and Other Biological Parameters of Immune Function in Older Adults," *International Journal of Psychiatry in Medicine,* 27 (1997): 233-250.
106—Rachel Dew and Harold Koenig, "Religious Involvement, the Serotonin Transporter Promoter Polymorphism, and Drug Use in Young Adults," *International Journal of Social Sciences,* 2 (2014): 1.

in the West at the dawn of the Industrial Revolution, when we were subject to harsh Darwinian selection. In evolutionary terms, this is "normal" and it is what was associated with those who survived brutal Darwinian conditions; in other words, those who carried a low percentage of (almost always bad) mutant genes. British psychologist Michael Woodley of Menie and his colleagues[107] note that once the Industrial Revolution hit in and Darwinian selection was substantially weakened, the mutant genes, that were previously cleansed every generation through Darwinian selection, resulting in a 40 percent child mortality rate, would no longer have been cleansed. Children with high levels of mutation would grow up and have children, and so would their children, resulting in a continuing build up of mutation. Mutant genes relating to the body are comorbid with those that relate to the mind; the brain being 84 percent of the genome. Accordingly, Woodley of Menie and his colleagues demonstrate, we find an increasing percentage of the population with genetic physical conditions and significantly genetic mental problems, such as autism, and these physical and mental problems do indeed tend to be co-morbid. Consistent with this, the same research team have presented evidence that our faces have become less symmetrical since pre-industrial times, strongly implying

107—Michael A. Woodley of Meinie, Matthew Sarraf, Radomir Pestow, and Heitor Fernandes, "Social Epistasis Amplifies the Fitness Costs of Deleterious Mutations, Engendering Rapid Fitness Decline Among Modernized Populations," *Evolutionary Psychological Science* 3 (2017): 181-191.

an increasing mutational load.[108] The authors suggest that in a context of growing mutational load what we would also expect to find is "spiteful mutations" that would lead to ways of thinking and acting that would be actively maladaptive under conditions of Darwinian selection, such as the desire to not have children, or the desire to strongly promote the interests of those who are very genetically distant from you over those of your own ethnic group. These genes would have been selected out under conditions of purifying Darwinian selection because they would generally correlate with mutant genes that would lead to a suboptimal immune system, for example. These mutants would help to undermine adaptive forms of behavior and adaptive group institutions, such that even those who did not carry the spiteful mutations would no longer be trained to act in a fully adaptive way and would begin to act in a maladaptive way.

We can understand how this would happen by comparing human society to a beehive. Woodley of Menie[109] argues that cooperation is so central to human, and bee, society that, in many crucial respects, these societies can be conceived of as kinds of superorganisms, with each individual playing a small but important role in the optimum functioning, and thus survival, of that superorganism. Every individual is part of a profoundly interconnected network wherein he relies on those with whom he interacts to behave in an adaptive fashion such

108—Michael A. Woodley of Menie and Heitor Fernandes, "The Secular Decline in General Intelligence From Decreasing Developmental Stability: Theoretical and Empirical Considerations," *Personality and Individual Differences*, 92 (2015): 194-199.
109—Dutton, *The Silent Rape Epidemic*, *op. cit.*.

that his own genes are expressed, phenotypically, in the same optimally adaptive fashion. It follows that if a feeder bee, in charge of feeding the larvae, has a mutation that causes her to feed the larvae at random, rather than according to an instinctively "normal" pattern, then she will damage the fitness of the entire hive, because there will be, for example, far too many queens and not enough workers.

Similarly, if a mutant human carries a spiteful mutation that makes him believe, and propagate the view that, life has no meaning, then his mutation will impact those who associate with him by causing, as with the bees, their environment to be different from that which their genes are optimally adapted to. This will mean that their genes will be expressed differently; sub-optimally. These non-carriers of spiteful mutations may be somewhat genetically predisposed to be Nihilists and abandon reproduction "because life is pointless" under only very specific and improbable conditions. If these conditions never occur, because the spiteful mutation carriers are not present in their society of purifying Darwinian selection, they will not develop those destructive behaviors. But once the spiteful mutant manifests, then he can undermine the societal culture, such as its religious rituals, and stop his genetically "normal" co-ethnics from being religious, meaning that his maladaptive worldview can spread like a plague even to those who lack the spiteful mutation. And, so, this single mutant can help to undermine the extent to which the group is optimally group-selected. If these spiteful mutants ascended to positions of power, they would be able to influence those who did not carry the spiteful mutations, by persuading them, for example, that women who are dedicated wives and mothers are "losers," to a very significant degree indeed, by heavily undermining the capacity of the group to engage in adaptive behavior and hold adaptive beliefs and so making even non-carriers maladaptive.[110] Woodley of Menie calls this the Social Epistasis Amplification Model.[111]

Continuing our exploration of this model, we would expect these "spiteful mutants" ascend to positions of power for a number of reasons. Firstly, we would expect it precisely because they are genetically maladaptive and their reaching such positions would be highly maladaptive for the group. In other words, they would yearn for power. Secondly, socioeconomic status has been shown to be about 70 percent genetic across generations,[112] meaning there is a strong degree to which social classes are genetic castes. High child mortality, and thus growing mutational load, would have commenced its decline, as the medical and other health-improving innovations of the Industrial Revolution took hold, firstly among those of higher socioeconomic status. Intelligence has been shown to be only very weakly associated with mutational load,[113] so we would expect the more intelligent social classes to concomitantly include a higher percentage of people carrying spiteful mutations. Thirdly, Gregory Clark has shown, through an analysis of rare surnames that reflected different Medieval social ranks, that between the Medieval period and 1950, there was selection for those from the

110—*Ibid.*
111—Woodley of Menie, Sarraf, Pestow, and Fernandes, "Social Epistasis Amplifies the Fitness Costs of Deleterious Mutations," *op. cit.*
112—Clark, *The Son Also Rises*, *op. cit.*
113—Woodley of Menie, Sarraf, Pestow, and Fernandes, "Social Epistasis Amplifies the Fitness Costs of Deleterious Mutations," *op. cit.*

"middling sort," something he also demonstrated in an analysis of sixteenth and seventeenth century English wills. Put simply, there was a process that Clark has termed the "Survival of the Richest." If English testators were divided in half, there richer 50 percent had a fertility advantage of 40 percent.[114] However, at the same time that this was happening, it is clear from Clark's research on surnames, that the nobility was contracting.[115] Noble families were dying out while the descendants of the wealthy plebeian class grew. This would likely reflect selection for intelligence, as the ability to accrue wealth is associated with intelligence. But, argues Clark, it also indicates that the values of the middle class, as distinct from the nobility and gentry, were helping them to survive. It has been shown that these were bourgeoisie values of competitive individualism[116] as well as more general burgher emphases on morality and education.[117] These stood in contrast to collectivist and even ethnocentric values associated with both the traditional nobility, with their sense of *noblesse oblige,* and the working class, who tended to be conservative, resistant to change and nationalistic. Clark charts the way in which those whose recent ancestors were "middling sort" gradually became part of the societal elite, substantially disempowering the traditional gentry and nobility by the second half of the twentieth century, as has been widely documented.

114—Clark, *A Farewell to Alms, op. cit.*
115—Clark, *The Son Also Rises, op. cit.*.
116— Henry R. French, "The Search for the 'Middle Sort of People' in England, 1600-1800," *The Historical Journal,* 43 (2000): 277-293.
117—Tony McEnery, *Swearing in English: Blasphemy, Purity and Power from 1586 to the Present* (London: Routledge, 2006).

Accordingly, we would expect "spiteful" ideas to emanate from the higher social classes. However, they would be most likely to emanate from the "middle class" or the New Elite, as they would be less likely to hold to the religiousness and conservatism associated with the old nobility. And we would expect, therefore, organs of power, such as the Church or academia, to increasingly reflect these "spiteful" ideas. An example of such a "spiteful" idea would be the way that Feminist leaders have managed to persuade so many females to essentially sacrifice any prospect of motherhood, inculcating them with the view that you are a failure if you are "just" a wife and mother. It occurred to me, therefore, that atheism, or, more specifically, the failure to collectively worship a moral god, could be just such a "spiteful mutation," most obviously because it seems to be negatively associated with fertility, health, and ethnocentrism.

The Mutant Says in His Heart: "There Is No God"

Myself, Guy Madison and American psychologist Curtis Dunkel decided to test this hypothesis in a study I entitled "The Mutant Says in His Heart 'There is No God.'"[118] If the theory were correct then religiousness, or this specific form of religiousness, would be negatively associated with markers of mutational load, and atheism, as well as other deviations such as belief in aliens and other paranormal

118—Dutton, Madison, and Dunkel, "The Mutant Says in His Heart, "There Is No God," *op. cit.*

phenomena lacking belief in a collectively worshiped moral god, would be positively associated with these markers. This is precisely what we found. Religiousness is negatively associated with autism, poor general health, and left-handedness, the latter being positively associated with physical and mental illness because it means that the brain has developed asymmetrically, implying mutational load. Belief in the paranormal was associated with poor physical health, schizophrenia, not being right handed, and being physically asymmetrical.[119] So, as predicted, the decline of religion, which is an adaptive trait under pre-modern conditions, is happening in the Western World in part due to the intense weakening of Darwinian selection. And this weakening has occurred due to our very high intelligence having brought about the fundamental breakthrough of the Industrial Revolution. Lack of traditional religiousness predicts not wanting children, so, in evolutionary terms, it is a road to oblivion.

Interestingly, Woodley of Menie and colleagues[120] highlighted John B. Calhoun's (1917-1995) famous "Mouse Utopia" experiment at the University of Maryland between 1968 and 1973[121] as eerily paralleling the Industrial Revolution and its consequences. The mice were placed in conditions where there was, essentially, no harsh Darwinian Selection. As with humans, there was an initial population explosion, but

119—Ray Blanchard, "Review and Theory of Handedness, Birth Order, and Homosexuality in Men," *Laterality* 13 (2008): 51-70.
120—Woodley of Menie, Sarraf, Pestow, and Fernandes, "Social Epistasis Amplifies the Fitness Costs of Deleterious Mutations," *op. cit.*
121—John B. Calhoun, "Death Squared: The Explosive Growth and Demise of a Mouse Population," *Proceedings of the Royal Society of Medicine* 66 (1973): 80-88.

then growth began to slow down as fewer and fewer mice had children. Eventually, the population plateaued and, as with developed countries where there is little or no immigration, it then went into decline. This was paralleled by the observation of very unusual forms of behavior. Females expelled their young from the nest too early so that they weren't sufficiently socialized to be able to deal with other mice. Females also became increasingly masculinized, aggressive and uninterested in breeding, even attacking their own young. Males, known as "the Beautiful Ones," became increasingly effeminate. They didn't fight for territory, had no interest in females, and spent all of their time grooming each other. Eventually all the mice were either "Beautiful Ones," masculinized females or socially inept. Consequently, no more mice were born, and the colony died out.

Woodley of Menie and colleagues show that the colony was nowhere close to being overcrowded when the population growth began to slow down. Thus, the simplest explanation is that mutant genes, causing maladaptive behavior in mice, were no longer being expunged from the population. This led to mutational meltdown. It is, of course, striking that one of the issues of our time is deviant sexuality and transsexuality in males and females, something associated with significantly genetic physical and mental illness and thus with high mutational load.[122] Due to their lower average intelligence, Islamic countries are under stronger conditions of Darwinian selection. Thus, clearly, to allow large numbers of young Muslim males to enter Western countries is akin to letting wild

122—Blanchard, "Review and Theory of Handedness, Birth Order, and Homosexuality in Men," *op. cit.*

animals into a zoo. Unlike mice, of course, humans have no John B. Calhoun to maintain their utopia. We have created our own utopia and while we are becoming more mutated we are also becoming less intelligent and being invaded by outsiders, so there will be no question of humanity dying out. As we argue in *At Our Wits' End,* the cycle of civilization will simply enter its winter, just as it did when Rome collapsed. Civilization will go backwards before, likely, rising again, as the cycle continues. Throughout history, civilizations have displayed a Spring, when they are highly religious, a Summer, an Autumn, when they are technologically advanced but also start to lose their religiousness, challenge traditions such as the Cult and aristocratic rule; and collapse into Winter and a new Dark Age.[123] It seems that we got further than ever before this time due to a combination of extremely high religiousness and an inherited religious taboo on contraception, meaning that the upper class didn't start seriously questioning their religion or developing contraception until *after* the Industrial Revolution had occurred.[124]

If our Atheism-Mutational Load Theory is correct, then we would also expect intelligence, beyond a certain level and thus weakly overall, to be negatively associated with religiousness and, as we have seen, this is indeed the case. Intelligence, in general, is associated with health, partly because intelligent people are more future-oriented and so eat more healthily. But, as we have already noted, intelligence is only adaptive up to a point, at least under

123—See Oswald Spengler, *The Decline of the West*, 2 vols., trans. Charles Francis Atkinson (Oxford: Oxford University Press (1926 [1918-1923]).

124—Dutton and Woodley of Menie, *At Our Wits' End, op. cit.*

the conditions of very low stress created by the Industrial Revolution. This is consistent with the evidence that very high intelligence predicts autism and not wanting to have children. Thus, the fact that intelligence predicts health but is negatively associated with religiousness, when religiousness is positively associated with health, does not in any way contradict the theory. Moreover, it has been found that the relationship between IQ and religiousness is not on general intelligence.[125] Accordingly, it may even be that the more intelligent were more religious, in the specific sense of the collective worship of a moral god, under Darwinian conditions, something evidenced by the fact that those of low status were inclined to worship or commune with, in addition to god, assorted invisible beings such as ancestors and sprites.[126]

Furthermore, until the Industrial Revolution, we were subject to selection for intelligence and selection for religiousness, but this was in a context in which we were also subject to considerable psychological stress, which was ameliorated, to some degree, by religiousness kicking in at a certain stress level. In this new context of very low stress, highly intelligent people, who would previously have also been religious to the extent of not using contraception and believing children were God's will, ceased to be religious and ceased to breed, rendering high intelligence (IQ above a certain level) manifestly

125—Edward Dutton, Jan te Nijenhuis, Guy Madison, Dimitri van der Linden, and Daniel Metzen, "The Myth of the Stupid Believer: The Negative Religiousness-IQ Nexus is Not on General Intelligence (*g*)," *Journal of Religion and Health* (2019), doi.org/10.1007/s10943-019-00926-3.

126—Dutton and Van der Linden, "Why is Intelligence Negatively Associated with Religiousness?," *op. cit.*

maladaptive in an evolutionary sense in a way that it was not under conditions of Darwinian selection. To put it another way, genes for high intelligence retrogressively became maladaptive and so we would expect them to be associated with forms of behavior which are more generally associated with markers of mutational load, such as atheism and not wanting to breed. Based on Woodley of Menie's Social Epistasis Amplification model, explored above, we would also expect intelligence, partly mediated by socioeconomic status, to be associated with desires that must surely indicate the presence of spiteful mutations, such as desiring the destruction of your own ethnic group via mass immigration. Consistent with this, intelligence predicts Leftist values in Western societies.[127] By contrast, religiousness continues to predict fertility among the natives. It continues to be adaptive.

Does Muhammad Have a Point?

It would seem to follow that what will save European people is to live in an environment that is harsher, less pleasant; less nice. To a certain extent, this will reintroduce conditions of Darwinian selection, it will make people more stressed, more instinctive, more religious and, as a consequence, more ethnocentric and more willing and able to fight for the survival of their own group. As our average

127—See Satoshi Kanazawa, *The Intelligence Paradox: Why the Intelligent Choice Isn't Always the Smart One* (Hoboken, NJ: John Wiley & Sons, 2012).

IQ declines, this will gradually happen anyway, but there is no reason to think that Western people will still exist in significant numbers by then, not least because they are being so heavily outbred by Muslims. Indeed, we would expect people from the Muslim world to be more adapted in a Darwinian sense because they are more religious and because, being from less developed societies, they have been subject to purifying Darwinian selection, for religiousness and mental and physical health, until more recently.

There is a battle for group selection. It is happening now. And European people are losing that battle because they are simply too intelligent and thereby too irreligious and too lacking in self-confidence; high self-esteem being a robust correlate of low IQ. A possible means, of course, by which Europeans could reduce their intelligence, and elevate their ethnocentrism, would be to convert to Islam. This is actually something that England once got very close to doing, at least if the non-contemporary source for this information is reliable.[128] King John Lackland (1166-1216) chronically mismanaged his kingdom during his disastrous reign between 1199 and 1216. Thus, he faced frequent revolts from his barons, who controlled areas of the country though were prepared, within reason, to be loyal to the anointed monarch. By 1212, England really was in a state of chaos. King John had been excommunicated by the Pope, Innocent III (1161-1216), some of his barons and parts of the country were in open revolt, and there was a serious possibility that the French would invade. According to the thirteenth century chronicler and Benedictine monk

128—See Ilan Shoval, *King John's Delegation to the Almohad Court (1212): Medieval Interreligious Interactions and Modern Historiography* (Turnhout, Belgium: Brephols Publishers, 2016).

Matthew Paris (c.1200-1259), in utter desperation, King John turned to Muhammad al-Nasir (reigned 1199-1213), the Caliph of Morocco, and offered to convert to Islam, converting his whole country with him, if the Caliph would help him. He sent emissaries to al-Nassir to plead his case, but the Caliph was so appalled by King John's groveling, as well as by his reputation as a tyrant, that he turned the delegation away.[129]

The French philosopher René Guénon (1886-1951) was certainly convinced that the only way the West could be saved was through conversion to Islam. In his book *The Crisis of the Modern World,*[130] published in French in 1946, Guénon argued, in effect, that the West was entering the winter of its civilization. It was exhausted, it had lost its sense of its own eternal importance, and it no longer really believed in its gods. It was going to collapse, as so many civilizations had before it, and be eclipsed by a younger civilization, brimming with the energy and confidence that you only acquire from believing that the gods really are guiding you towards glory. More recently, some philosophers, such as Sir Roger Scruton,[131] and even mainstream journalists are increasingly commenting on this, noting that there is a kind of spiritual crisis in the West, "The feeling that the story has run out."[132]

129—John V. Tolan, *Saracens: Islam in the Medieval European Imagination* (New York: Columbia University Press, 2002), 183; Aomar Boum and Thomas K. Park, *Historical Dictionary of Morocco* (Lanham, MD: Rowman and Littlefield, 2016), 209.
130—Réne Guénon, *The Crisis of the Modern World* (Hillsdale, NY: Sophia Perennis, 2004 [1946]).
131—Roger Scruton, *Modern Culture* (London: Continuum, 2000).
132—Douglas Murray, *The Strange Death of Europe: Immigration, Identity, Islam* (London: Bloomsbury, 2007).

Guénon maintained that civilizations that are flourishing and confident are invariably religious. They genuinely believe in the existence of a transcendent realm that is the ultimate source of all authority. In fact, they don't even "believe" it. They *know* that there are some things that are perennial, that are permanent, and there can only be truth and understanding with reference to these things; with reference to a metaphysical reality. This is the essence of the philosophical school known as Traditionalism.[133] If you don't know of this reality, then your civilization is doomed to collapse. It is doomed to collapse because, without a sense of eternal significance, it is difficult to defend civilization. Indeed, it is doomed to collapse because, without belief in a metaphysical reality, there can be no objective sense of truth, science will be debased and a more "scientific" society will develop superior weapons and triumph over you. This may seem contradictory, but it can be argued that the only way scientific endeavor can continue is if you believe in the fundamental importance of knowing the truth. American political scientist Charles Murray has observed that many nineteenth century genius scientists were so called Neo-Thomists, a reference to the Medieval Scholastic philosophy of St Thomas Aquinas (1224-1274). They believed it was their divinely-inspired duty to uncover the truth about God's creation and that it was blasphemous to lie about it. Ancient Greek science developed out of the pagan desire to attain the

133—Mark Sedgwick, *Against the Modern World: Traditionalism and the Secret Intellectual History of the Twentieth Century* (Oxford: Oxford University Press, 2004).

knowledge held by the gods and thus reach the status of a god: it assumed an objective metaphysical reality which verified objective truth.[134] Islamic science was influenced by the Neo-Platonic belief in an objective metaphysical reality of Truth, a World of Forms, as well as by the Prophet Muhammad's exhortation to understand Allah's creation. Objective truth, it might be argued, presupposes something eternal. Without this belief, "truth" is no longer sacred and science can easily become corrupted. Indeed, it might be argued that the increasing suppression of "controversial science," such as in 2019 when Noah Carl was sacked by St Edmund's College, Cambridge, for his research on areas relating to race,[135] reflects the increasing influence of atheists and especially postmodernists, who see truth, on some level, as relative and reducible to "power," in science. As Bruce Charlton has put it:

> *The root of the problem is, I believe, that modern scientists have progressively abandoned their belief in the reality of truth and the conviction that the life of a scientist must be characterized by truthfulness in all things great and small. Nowadays, scientists are embarrassed to talk or write about truth: indeed, modern scientists regard 'truth talk' as naïve, amateurish, hypocritical and probably manipulative.*

134—Benoist, *On Being a Pagan, op. cit.*
135—*Quillette*, "Noah Carl: An Update on the Young Scholar Fired by a Cambridge College for Thoughtcrime," May 28, 2019, https://quillette.com/2019/05/28/noah-carl-an-update-on-the-young-scholar-fired-by-a-cambridge-college-for-thoughtcrime/ (accessed August 15, 2019).

> *Yet until about fifty years ago, scientists were raised and lived in a culture so permeated with religious understandings and transcendental values that direct personal derivation from Church teachings was hardly necessary. Even when they became atheists, scientists remained religious about truth. However, the combination of atheism with belief in the reality and importance of truth seems to have been unstable and unsustainable and is now rare.*[136]

In other words, even if scientists thought they were atheists they adhered to what English theologian Edward Bailey (1935-2015) has called an "implicit religion":[137] A *de facto* religious belief discernible from analyzing their words that is something like Neo-Platonism.

For Rene Guénon, as with the discipline of Religious Studies which we discussed earlier, religions have in common this certainty that there exists such a perennial realm. It follows that if, through some kind of effortful control or self-hypnotism, we can once again compel ourselves to accept this then the rejuvenation of our dying civilization will follow, presumably because being genuinely religious, we will once again be ethnocentric. This is where traditionalists diverge from conservatives such as the English philosopher Sir Roger Scruton who argues that we should live "as if" our lives have eternal importance.[138] You don't live "as if" our lives have eternal importance. You *know* they

---

136—Bruce G. Charlton, "Scientists Need to Rediscover Truth, *Church Times*, June 19, 2009.
137—Edward Bailey, *Implicit Religion* (Milton Keynes: Middlesex University Press, 1997).
138—Scruton, *Modern Culture, op. cit.*

do. But how can we persuade ourselves to re-embrace this perennial philosophy that is religiosity?

Usually religious fervor occurs through some kind of religious experience, in which you somehow feel a sense of awe and simply know that God and the metaphysical world is real.[139] This experience tends to be navigated via your society's traditional religious symbols. But, argues Guénon, the West has lost touch with these. Christianity, for example, has been criticized, ridiculed, dissected and pulled apart; for many Westerners it is little different from maintaining traditions such as Father Christmas. If Christianity has a purpose, it is as part of a kind of hollow traditionalism that at least gives life structure, and the feeling that we're doing what our ancestors have done, at times of crisis, such as funerals. The Rev. Anthony Freeman, a now defrocked Church of England vicar, went so far, in 1993, as to openly state that this was the essence of his Christianity. His church was a bit like a local society called the Sealed Knot, which re-enacted battles from England's Civil War. It was engaging in rituals that our ancestors had and so creating a sense of superficial permanence. And he certainly didn't believe in God. To claim God existed was like saying "the wind is green."[140]

Guénon maintains that in the East, by contrast, living aspects of traditional civilization have been preserved. In the East, people still know that the transcendent world is real and, accordingly, their religions have managed to preserve a sense of mystery and awe. Accordingly, initiation

139—See Lewis Rambo, *Understanding Religious Conversion.* New Haven: Yale University Press (1993).

140—Anthony Freeman, *God In Us: A Case for Christian Humanism* (London: SCM, 1993).

into an Eastern religion will far better induce our latent religious feelings than will some attempt to convert to the moribund religion that is Western Christianity. Initiates will be galvanized and, with the zeal of the convert, they will be able to recruit others.[141] For Guénon the selected Eastern religion needs to be relatively similar to Christianity, due to the significance of Christianity to our culture. Thus, the best option is Islam. In 1910, Guénon was initiated into a Moroccan Sufi order. He adopted the Islamic name Abd al-Wāḥid Yaḥyā and eventually moved to Egypt, where he died in 1951. His last word was "Allah."[142] The Italian baron and philosopher Julius Evola (1898-1974) was influenced by Guénon, but felt that much of the East had also lost contact with the traditional world[143] and that, accordingly, Europeans were better off attempting to get in touch with the latent religious feelings they still had for Europe's traditional pagan beliefs[144] or possibly even immersing themselves in a continuous "war against the modern world," with religious transcendence reached during the heat of

141—Julius Evola, "René Guénon," *East and West*, vol. 4, no. 4 (1954): 255–58; *Counter Currents*, trans. Greg Johnson, *Counter-Currents*, December 2, 2010, https://www.counter-currents.com/2010/12/rene-guenon-east-west/ (accessed August 15, 2019).
142—Paul Chacornac, *The Simple Life of René Guénon*, trans. Ccil Beshell, Hillsdale, NY: Sophia Perennis, 2005).
143—Evola, "Réné Guénon," *op. cit.*
144—Julius Evola, *Revolt Against the Modern World: Politics, Religion, and Social Order in the Kali Yuga.* trans. Guido Stucco (New York: Simon & Schuster, 1995 [1934]). Julius Evola, *Ride the Tiger: A Survival Manual for Aristocrats of the Soul*, trans. Joscelyn Godwin & Constnace Fontana. (New York: Simon & Schuster, 2003 [1961]).

this battle.[145] Another possibility may be Westerners embracing the Orthodox Church, insomuch as a remnant genuine attachment to Christianity remains in the West. This church appears to be relatively resistant to influence from secular ideologies, possibly due to its strong focus on ritual over theology: "The education of the clergy is restricted to the training of ritualistic technicians."[146] It is heavily syncretized with aspects of European paganism, and may be attractive precisely because it is "Eastern": "This is an essentially Oriental form of Worship, gorgeous and *ritualistic*, full of mysticism and allegory, not perhaps easily understood by the Protestant, who strives after God mainly through his intellect."[147] Curious and high in Openness as Westerners are, they have for centuries been attracted to and fascinated by the "mysterious East" for millennia, whether to Eastern mystery cults, Christianity or religions developed in India.[148] Accordingly, the Orthodox Church, unlike other branches of Christianity, may possibly still possess a sense of religious mystery, of *mysterium, tremendem et fascinans,* for Westerners, allowing us to become a focus for their remnant or latent Christianity.

145—Evola, *Ride the Tiger*, *op. cit.*; Julius Evola, *Metaphysics of War: Battle, Victory and Death in the World of Tradition* (London: Arktos Publishing, 2011[1996]). See Mark Sedgwick, *Against the Modern World: Traditionalism and the Secret Intellectual History of the Twentieth Century* (Oxford: Oxford University Press, 2004).

146—William Stroyen, *Communist Russia and the Russian Orthodox Church, 1943-1962* (Washington, DC: Catholic University Press of America, 1967), 20.

147—Constantine Callinicos, *The Greek Orthodox Church* (London: Longmans, Green & Co, 1918), 28.

148—Richard King, "Orientalism and the Modern Myth of 'Hinduism.'" *Numen,* 46 (1999): 146-185.

Out of the Mouth of Babes . . .

Recently, the Canadian psychologist-guru Jordan B. Peterson[149] has popularized his own philosophy, that has some points in common with Traditionalism. He presents a kind of eclectic religion whereby all religious traditions are regarded as portholes into the world of myth and entering these can somehow give our lives meaning and direction, in line with his inspiration Carl Jung (1875-1961). He also insists that, deep down, we all know that we believe in God really.

> *You might object, "But I'm an atheist!" No, you're not . . . You're simply not an atheist in your actions, and it is your actions that most accurately reflect your deepest beliefs – that are implicit, embedded in your being, underneath your conscious apprehensions and articulable attitudes and surface-level knowledge. You can only find out what you actually believe (Rather than what you think you believe) by watching how you act. You simply don't know what you believe before that. You are too complex to understand yourself.*[150]

This is an excellent and well-articulated point. But the obvious problem with Peterson's manifesto is that it is unclear how it helps to revive Western civilization. Peterson has dismissed identification with your ethnic group, in other words "ethnocentrism," as a sign of

149—Jordan B. Peterson, *12 Rules For Life: An Antidote to Chaos* (London: Allen Lane, 2018).
150—*Ibid.*, 103.

being a pathological loser.[151] In other words, unlike Traditionalists who will simply allow traditional values, such as ethnocentrism, to naturally flow from religious revival, if you are initiated into Peterson's *ashram*, you must accept that any interest in your genetic interests is an indication of a childish, simple mind. It is precisely this kind of thinking that will ensure that the West is absolutely trounced in the battle of group selection. Based on current trends, Europeans will at worst die out completely and, at best, be conquered by Islam, interbreeding with their new masters. On the positive side, this will broaden the Islamic gene pool, elevate the new group's average intelligence and so perhaps mean that *Eurabia* is less backward than it might otherwise be. But average national IQ predicts every aspect of civilization: sanitation, the rule of law, trust, education, wealth, political stability, health, life expectancy, even happiness.[152] *Eurabia* will be a shadow of what it could have been.

Islam makes you stupid. But, in doing so, it makes you strongly ethnocentric and, if balanced with an optimum level of intelligence, far more likely to triumph in our on-going evolutionary saga. We began this exploration with the words of Winston Churchill. Much criticism has been leveled against him. He may well have

---

151—Jordan B. Peterson, "On the So-Called 'Jewish Question.'" *JordanPeterson.com,* https://jordanbpeterson.com/psychology/on-the-so-called-jewish-question/ (accessed August 15, 2019); see also Richard Spencer and Peter Gast, "The Peterson Pathology," *Radix Journal Podcast*, https://www.spreaker.com/user/altright/jordan-peterson-response (accessed August 15, 2019).
152—Lynn and Vanhanen, *Intelligence, op. cit.*

taken the West into a suicidal and unnecessary war,[153] though many disagree. But he was certainly brilliant with words, particularly in his hypnotic wartime speeches:

> *We have before us an ordeal of the most grievous kind. We have before us many, many long months of struggle and of suffering. You ask, what is our policy? I will say: It is to wage war, by sea, land, and air, with all our might and with all the strength that God can give us; to wage war against a monstrous tyranny never surpassed in the dark, lamentable catalogue of human crime. That is our policy. You ask, what is our aim? I can answer in one word: It is victory, victory at all costs, victory in spite of all terror, victory, however long and hard the road may be.*[154]

In the struggle for the future of the world, victory for the West is more likely if Europeans are more religious, less cerebral, and possessed by feelings of camaraderie and destiny—in other words, if they become more like those against whom they are competing. In this regard, Islam is right about many things . . . many, many things.

153—See Patrick J. Buchanan, *Hitler, Churchill, and the Unnecessary War: How Britain Lost Its Empire and the West Lost the World* (New York: Crown Publishers, 2008).

154—Winston Leonard Spencer-Churchill, "First Speech as Prime Minister to House of Commons," May 13, 1940, "Blood, Toil, Tears, and Sweat," *International Churchill Society*, https://winstonchurchill.org/resources/speeches/1940-the-finest-hour/blood-toil-tears-and-sweat-2/ (accessed August 15, 2019).

INDEX

#

A

B

C

D

E

F

G

M

N

O

P

Q

R

S

T

U

V

W

Y

About the Author

EDWARD DUTTON is Editor at Washington Summit Publishers and an independent researcher based in Finland. Born in London in 1980, Dutton read Theology at Durham University before completing a PhD in Religious Studies at Aberdeen University in 2006. His research there was developed into his first book, *Meeting Jesus at University: Rites of Passage and Student Evangelicals*. He was made Docent of the Anthropology of Religion and Finnish Culture at Oulu University in 2011. In 2012, Dutton made the move to evolutionary psychology and has never looked back. Dutton has published in leading psychology journals including *Intelligence*, *Personality and Individual Differences*, and the *Journal of Biosocial Science*. Among other positions, Dutton has been a guest researcher at Leiden University in the Netherlands and Umeå University in Sweden, visiting professor of the Anthropology of Religion at Riga Stradins University in Latvia, and academic consultant to a research group at King Saud University in Riyadh, Saudi Arabia. Dutton's research has been reported on worldwide, by the *Daily Telegraph*, *The Sun*, *Le Monde*, *Newsweek*, among other publications, and his writings have been translated into Russian, Portuguese, Spanish, and Czech. Dutton can be found online at his award-winning channel *The Jolly Heretic*. He is married to a Finn, has two young children, and enjoys Indian cooking and genealogy.

CPSIA information can be obtained
at www.ICGtesting.com
Printed in the USA
LVHW040509100221
678896LV00005B/345